Machine Learning: Foundations, Methodologies, and Applications

Series Editors

Kay Chen Tan, Department of Computing, Hong Kong Polytechnic University, Hong Kong, China

Dacheng Tao, University of Technology, Sydney, Australia

Books published in this series focus on the theory and computational foundations, advanced methodologies and practical applications of machine learning, ideally combining mathematically rigorous treatments of a contemporary topics in machine learning with specific illustrations in relevant algorithm designs and demonstrations in real-world applications. The intended readership includes research students and researchers in computer science, computer engineering, electrical engineering, data science, and related areas seeking a convenient medium to track the progresses made in the foundations, methodologies, and applications of machine learning.

Topics considered include all areas of machine learning, including but not limited to:

- Decision tree
- Artificial neural networks
- Kernel learning
- Bayesian learning
- Ensemble methods
- Dimension reduction and metric learning
- Reinforcement learning
- Meta learning and learning to learn
- Imitation learning
- Computational learning theory
- Probabilistic graphical models
- Transfer learning
- Multi-view and multi-task learning
- Graph neural networks
- Generative adversarial networks
- Federated learning

This series includes monographs, introductory, and advanced textbooks, and state-of-the-art collections. Furthermore, it supports Open Access publication mode.

Yaochu Jin · Hangyu Zhu · Jinjin Xu · Yang Chen

Federated Learning

Fundamentals and Advances

 Springer

Yaochu Jin
Faculty of Technology
Bielefeld University
Bielefeld, Germany

Jinjin Xu
Intelligent Perception and Interaction
Research Department
OPPO Research Institute
Shanghai, China

Hangyu Zhu
Department of Artificial Intelligence
and Computer Science
Jiangnan University
Wuxi, China

Yang Chen
School of Electrical Engineering
China University of Mining and Technology
Xuzhou, China

ISSN 2730-9908   ISSN 2730-9916   (electronic)
Machine Learning: Foundations, Methodologies, and Applications
ISBN 978-981-19-7085-6   ISBN 978-981-19-7083-2   (eBook)
https://doi.org/10.1007/978-981-19-7083-2

This Springer imprint is published by the registered company Springer Nature Singapore Pte Ltd.
The registered company address is: 152 Beach Road, #21-01/04 Gateway East, Singapore 189721,
Singapore

Preface

I heard the terminology "federated learning" for the first time when I was listening to a talk given by Dr. Catherine Huang from Intel at a workshop of the IEEE Symposium Series on Computational Intelligence in December 2016 in Athens, Greece. I was fascinated by the idea of federated learning and immediately recognized the paramount importance of preserving data privacy, since deep learning had been increasingly relying on the collection of a large amount of data, either from industrial processes or from human everyday life.

Federated learning has now become a popular research area in machine learning and received increased interest from both industry and government. Actually already in April 2016, the European Union adopted the General Data Protection Regulation, which is the most strict law on data protection and privacy in the European Union and the European Economic Area and took effective in May 2018. Consequently, data privacy and security has become indispensable for practical applications of machine learning and artificial intelligence, together with other importance requirements, Including explainability, fairness, accountability, robustness and reliability.

This book presents a compact yet comprehensive and updated coverage of fundamentals and recent advances in federated learning. The book is self-contained and suited for both postgraduate students, researchers, and industrial practitioners. Chapter 1 starts with an introduction to the most popular neural network models and gradient-based learning algorithms, evolutionary algorithms, evolutionary optimization, and multi-objective evolutionary learning. Then, three main privacy-preserving computing techniques, including secure multi-party computation, differential privacy, and homomorphic encryption are described. Finally, the basics of federated learning, including a category of federated learning algorithms, the vanilla federated averaging algorithm, federated transfer learning, and main challenges in federated learning over non independent and identically distributed data are presented. Chapter 2 focuses on communication efficient federated learning, which aims to reduce the communication costs in federated learning without deteriorating the learning performance. Two algorithms for reducing communication overhead in federated learning are detailed. The first algorithm takes advantage of the fact that shallow layers in a deep neural network are responsible for learning general features across different classes while deep layers take care of class-specific features. Consequently, the deep layers can be updated less frequently than the deep layers, thereby reducing the number of parameters that need to be uploaded and downloaded. The second algorithm adopts a completely different approach, which dramatically decreases the communication load by converting the weights in real numbers into ternary values. Here, the key question is how to maintain the learning performance after ternary compression of the weights, which is achieved by training a real-valued co-efficient for each layer. Interestingly, we are able to prove theoretically that model divergence can even be mitigated by introducing ternary quantization of the weights, in particular when the data is not independent and identically distributed. While layer-wise synchronous weight update and ternary compression described in Chap. 2 can effectively lower the communication cost, Chap. 3 addresses communication cost by simultaneously maximizing the learning performance and minimizing the model complexity (e.g., the number of parameters in the model) using a multi-objective evolutionary algorithm. To go a step further, a real-time multi-objective evolutionary federated learning algorithm is given, which searches for optimal neural architectures by maximizing the performance, minimizing the model complexity, and minimizing the computational performance (indicated by floating point operations per second) at the same time. To make it possible for real-time neural architecture search, a strategy that searches for subnetworks by sampling a supernet, and reducing computational costs by sampling clients is introduced. To enhance the security level of federated learning, Chap. 4 elaborates two secure federated learning algorithms by integrating homomorphic encryption and differential privacy techniques with federated learning. The first algorithm is based on a horizontal federated learning, in which a distributed encryption algorithm is applied to the weights to be uploaded on top of ternary quantization. By contrast, the second algorithm is meant for vertical federated learning based on decision trees, in which secure node split and construction are developed based on homomorphic encryption and predicted labels are aggregated on the basis of partial differential privacy. This algorithm does not assume that all labels are stored on one client, making it more practical for real-world applications. The book is concluded by Chap. 5, providing a summary of the presented algorithms, and an outlook of future research.

The two algorithms presented in Chap. 2 were developed by two visiting Ph.D. students I hosted at University of Surrey, Jinjin Xu and Yang Cheng. The two multi-objective evolutionary federated learning algorithms were proposed by my previous Ph.D. student, Hangyu Zhu. Finally, the two secure federated learning algorithms were designed by Hangyu Zhu, in collaboration with Dr. Kaitai Liang, and his Ph.D. student, RuiWang, who were with the Department of Computer Science, University of Surrey, and are now with the Department of Intelligent Systems, Delft University of Technology, The Netherlands. I would like to take this opportunity thank to Kaitai and Rui for their contributions to Chap. 4.

Finally, I would like to acknowledge that my research is funded by an Alexander von Humboldt Professorship for Artificial Intelligence endowed by the German Federal Ministry of Education and Research.

Bielefeld, Germany   Yaochu Jin
August 2022

Contents

Chapter 1
Introduction

Abstract This chapter introduces the background knowledge of the book, including the most widely used artificial neural network models and decision trees, gradient based learning methods, evolutionary algorithms and their applications to single- and multi-objective machine learning, traditional privacy-preserving computing methods such as multi-party secure computation, differential privacy, and homomorphic encryption, and the federated learning paradigm for privacy-preserving machine learning. An overview of horizontal and vertical federated learning, together with a description of the basic federated learning algorithm, known as federated averaging, is presented, before knowledge transfer in federated learning is briefly explained. Finally, the main challenges of federated learning over non independent and identically distributed data are discussed in detail.

1.1 Artificial Neural Networks and Deep Learning

1.1.1 A Brief History of Artificial Intelligence

The terminology of artificial intelligence was formally suggested in a proposal of a summer workshop held in Dartmouth 1955, although the idea to create a programmable machine can be traced back to the 19th century [1]. The first mathematical model proposed by McCulloch and Pitts in 1943 [2], known as the McCulloch-Pitts model, simulates the function of a single neuron. In 1949, the Hebb Law was proposed by Hebb [3], which states that neurons that fire together wire together, meaning that a connection between two neurons will become stronger if they activate simultaneously, and the connection will weaken if they activate at different times. Many unsupervised learning algorithms were based the Hebb Law in principle. The earliest functional neuron model is the perceptron proposed by Rosenblatt in 1958 [4], which can separate linearly separable patterns.

During 1945 and 1947, Alan Turing, a pioneer of both computer science and artificial intelligence, worked on the design of the Automatic Computing Engine, which is widely recognized as the first stored-program computer. In 1950, Turing proposed his most well known test to define if a machine is intelligent, called Turing test [5]. The basic idea behind the Turing test is that a machine can be seen as intelligent if a human interrogator is not able to distinguish a machine from a human being through conversation. The term machine learning was first proposed by Arthur Samuel in 1959, which was defined as a research field that enables computers to learn without being explicitly programmed. A large body of research on artificial intelligence was carried out after the Dartmouth workshop, including the development of the first natural language processing computer program ELIZA [6] and the first industrial robot Unimate [7]. It should be mentioned that two important research areas of artificial intelligence, namely fuzzy sets and fuzzy systems [8] that simulate human reasoning, and evolutionary computation [9] that simulates natural evolution, were also developed during the 1960s.

The first 'winter' of artificial intelligence started in the beginning of 1970s after Minsky and Papert published a book analyzing the limitations of perceptrons [10]. Nevertheless, several prominent advances were made during the 1970 and 1980s, including the development of many successful expert systems, proposal of an early version of the error back-propagation learning algorithm for a neural network model containing one hidden layer that can solve the XOR problem [11], and genetic algorithms [12] as well as population based evolution strategies [13].

Several breakthroughs were achieved in the 1980s. In 1982, Kohonen proposed the self-organizing maps [14], which is one most important unsupervised learning algorithms, while reinforcement learning was suggested in 1983 [15]. The Hopfield neural network, which is a recurrent neural network model, remarked the start of second boom of the artificial intelligence research, which was culminated by the publication of the seminal paper [16] in 1986, in which the error back-propagating algorithm was proposed for training multiple layer perceptrons, enabling artificial neural networks containing one or two hidden layers to solve many linearly nonseparable classification problems and nonlinear regression problems.

Many other learning algorithms and neural network models have also been reported during the 1980s. In 1983, the Bienenstock-Cooper-Munro (BCM) neural plasticity rule was published [17], which can be seen as an extension of the Hebbian rule and has become a powerful unsupervised learning algorithm. After more than one decade, another important unsupervised learning rule, called spike-timing dependent plasticity was developed in 1998 [18], which has become one popular unsupervised learning algorithm for spiking neural networks proposed in 1997 by Wolfgang Maas [19]. Very different from other artificial neural networks that are based on continuous analog signals, spiking neural networks use the timing and frequency of the spikes or pulses for information processing, which is biologically more plausible since all biological nervous systems, including the human brain, also rely on spikes. In the meantime, Neurocognitron [20], which was based on the architecture of the human visual system was suggested by Fukushima in 1983, and which can be seen as the predecessor of the convolutional neural network [21] proposed in 1998, which is currently the most popular neural network model for solving computer vision problems. Several other important pieces of research on artificial intelligence have also been proposed in the 1990s, e.g., the support vector machine [22], and the long short-term memory, a more powerful recurrent neural network model [23]. Interestingly, support vector machines and other statistically learning algorithms outperformed those based on artificial neural networks, resulting in the second 'winter' of the artificial intelligence research.

There were also lots of interesting new developments outside the field of neural networks and machine learning during the 1990s. Specifically, new theories and methodologies were developed in fuzzy systems, including the introduction of type-2 fuzzy sets and a wide range of successful applications of fuzzy control and fuzzy decision-making. Besides, evolutionary computation has also become a popular research field, in which not only new paradigms such as genetic programming was proposed, but several other population based meta-heuristics were also proposed, including particle swarm optimization, ant colony optimization, and differential evolution [24]. Evolutionary algorithms and other metaheuristics have been shown to be effective in solving complex optimization problems, in particular multi-objective optimization problems, dynamic optimization problems, and data-driven optimization problems.

Like in the first winter of artificial intelligence, many researchers continued working in the field of neuronal networks, fuzzy systems and evolutionary computation, which was called computational intelligence, instead of artificial intelligence, mainly because artificial intelligence became unwelcome in the research community. Many popular emerging research topics in artificial intelligence nowadays have already been studied in the 1990s and early 2000s, before the third wave of artificial intelligence. These include interpretability of trained neural networks and fuzzy systems, robust machine learning, multi-objective machine learning and structure optimization of neural networks, now popularly known as neural architecture search.

The third resurgence of artificial intelligence started in 2007 when Hinton published a paper on effective training of artificial neural networks consisting of many large hidden layers [25], now widely known as deep neural networks. The training of deep neural networks was considerably accelerated by using graphic processing units (GPUs), making it possible to effectively train deep neural networks containing tens or even hundreds of hidden layers on huge datasets such as ImageNet by taking advantage of the immense computational resources available nowadays. A revolutionary success was achieved by deep learning when a computer go player algorithm called AlphaGo on the basis of deep reinforcement learning and Monte Carlo tree search defeated a human world champion.

To date, deep learning has achieved tremendous successes that were unimaginable years ago. Deep learning has demonstrated unique power in solving almost all single problems in science and technology, ranging from face recognition, game playing, to healthcare, natural language processing, protein folding and drug discovery. Many very powerful deep neural network models have been proposed, such as autoencoder, variational autoencoder, generative adversarial networks, and transformer, just to name a few. In addition, new research paradigms such as transfer learning, few shots learning, and self-supervised learning have been developed to handle various problems, in particular deep learning in the presence of data paucity.

Despite the above successes deep learning has achieved, many technical, ethical, and social issues remain or have arisen. Technically, it is still a challenging issue to enable machine learning models to learn multiple tasks on small data, let alone to learn autonomously. Trustworthiness and fairness of artificial intelligence become increasingly concerning as it is more and more widely used in critical, human, and societal systems. Trustworthiness typically include explanability and or transparency, safety, reliability and robustness, privacy preservation, respect of human values, green, and accountability.

1.1.2   Multi-layer Perceptrons

A *multi-layer perceptron* (MLP) [26–28] often represent a fully connected feed-forward artificial neural network (ANN). And it consists of three types of layers: the input layer, output layer and hidden layers, each of which contains several neurons named perceptrons [29]. Before discussing about MLP neural networks, simpler perceptrons will be explained at first.

1.1.2.1   Perceptrons

The general structure of a perceptron is shown in Fig. 1.1, where $\mathbf{x} = (x_1, x_2, \ldots, x_n)$ is a n-length input feature vector, b is the bias and $\mathbf{w} = (w_1, w_2, \ldots, w_n)$ is the corresponding weight vector of the input features.

For feed-forward propagation, a perceptron receives a weighted sum of the input features together with the bias as the input z computed by the following Eq. (1.1):

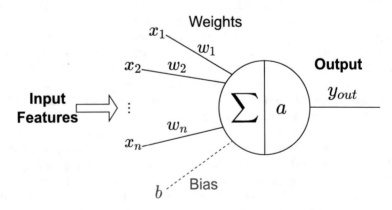

Fig. 1.1 An example of perceptron

Fig. 1.2 An example of the
step function when the
threshold t is 0

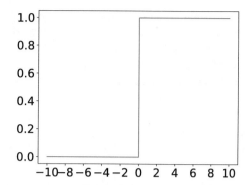

$$z = \sum_{i=1}^{n} x_i w_i = \mathbf{x}^T \mathbf{w} \tag{1.1}$$

The computed sum z is then passed through an activation function a for nonlinear transformation, refer to Fig. 1.1. And the step function is always selected to be the activation function of the perceptrons whose outputs are calculated according to Eq. (1.2):

$$y_{out} = a(z) = \begin{cases} 1, \ z > t. \\ 0, \ z < t. \end{cases} \tag{1.2}$$

where t is the threshold. A simple example of the step function when $t = 0$ is shown in Fig. 1.2. If the weighted sum z is larger than the threshold t, the perceptron in Fig. 1.1 will output $y_{out} = 1$; otherwise, the output y_{out} will become 0. Note that the step function can be replaced by other activation functions or even be removed for specific learning tasks.

1.1.2.2   Activation Function

For a typical supervised learning problem, we redefine the outputs of the perceptron for any input vector $\mathbf{x}$ as model prediction $\hat{y}_{out}$ and the corresponding data label as y_{out}. In this case, training perceptrons can be converted into minimizing the constructed loss function representing the distance between the prediction $\hat{y}_{out}$ and the actual label y_{out}, which is conducted by optimizing the weights and bias (often called model parameters) of the perceptrons.

The gradient based optimization methods are commonly used in training perceptrons. The core idea of this kind of approach is to let the model parameters recursively subtract their corresponding product of the gradients and the learning rate until the loss function converges. And calculating the gradients of the model parameters with the chain rule [30] requires the constructed loss function to be continuous and differentiable. Therefore, the above mentioned step function is usually approximated by the sigmoid function, making it differentiable for the gradient based training.

It is clear to see that the curve of sigmoid function shown in Fig. 1.3(a) has a similar shape as that of step function shown in Fig. 1.2. These two functions also have the same output range, and the sigmoid function can be seen as the smoothed and continuous version of the step function. The sigmoid function is described in Eq. (1.3):

$$\hat{y}_{out} = \sigma(z) = \frac{1}{1 + e^{-z}} \tag{1.3}$$

And the derivative of the sigmoid function with respect to its input z (weighted sum of the input data features) can be easily calculated in Eq. (1.4):

$$\begin{aligned} \frac{\partial \sigma(z)}{\partial z} &= -\frac{1}{(1 + e^{-z})^2} \cdot (-e^{-z}) \\ &= \sigma(z)(1 - \sigma(z)) \end{aligned} \tag{1.4}$$

From Fig. 1.3(b), the largest gradient of the sigmoid function is located at the central part of the curve and it tends to approach 0 when the absolute value of the input z approaches to infinity. That is the reason why normalization techniques [31–33] are often used in modern machine learning tasks.

Furthermore, the sigmoid function may cause gradient vanishing especially for MLPs with a large number of hidden layers. This is because the gradients of MLP neural networks are calculated using the error back propagation (to be discussed later in detail) by computing the derivative from the last output layer to the input layer. And according to the chain rule, the gradients of shallower layers (closer to the input layer) are derived by multiplying the calculated derivatives of deeper layers (closer to the output layer). Consequently, when calculating, for example, the gradients of the shallower layer for a n-hidden-layer MLP using the sigmoid activation function, n small derivatives (the maximum value is 0.25 shown in Fig. 1.3(b) are multiplied together, making the resulting gradients extremely small. In this case, the gradients

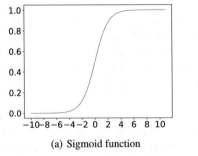
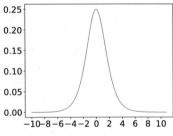

(a) Sigmoid function   (b) Derivative of the sigmoid function

Fig. 1.3 An example of the sigmoid function and its corresponding derivative

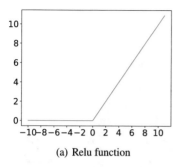

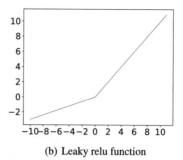

(a) Relu function   (b) Leaky relu function

Fig. 1.4 ReLU and leaky ReLU functions

are too small to have negligible influence on the parameter updates during the training period.

To overcome this issue, rectified linear unit (ReLU) or leaky ReLU is instead selected to be the activation function as described in Eq. (1.5):

$$\hat{y}_{out} = a(z) = \begin{cases} z, & z \geq 0. \\ az, & z < 0. \end{cases} \tag{1.5}$$

where a is real-valued hyperparameter with a range of $[0, 1)$, and above equation becomes ReLU function if $a = 0$; otherwise it called is leaky ReLU. And the corresponding plot is also shown in Fig. 1.4.

It is easy to find that the derivatives of both ReLU and leaky ReLU are 1 if the input $z > 0$, thus, effectively avoiding the gradient vanishing issue of multiplying derivatives layer by layer compared to the sigmoid function. In addition, the ReLU function is computationally efficient, where only comparison, addition and multiplication operations are involved during model training. Therefore, ReLU has become the most popular activation function for modern deep neural networks [34].

1.1.2.3   Model Structure

A perceptron without any connection (the node represented by a big circle in Fig. 1.1) is acted as a basic building element called a neuron in MLP neural networks. Roughly speaking, MLPs are a brunch of neurons connected together with a multi-layer structure as shown in Fig. 1.5, where circles in solid lines are neurons containing the activation function and circles in dashed lines represent biases with weight connections equal to 1.

When a training data passes through the MLP model from the input layer to the output layer, exactly the same forward computation is performed as a perceptron does for each neuron of each layer. And the mathematical formulation for the output of each layer is described as follows:

Bias

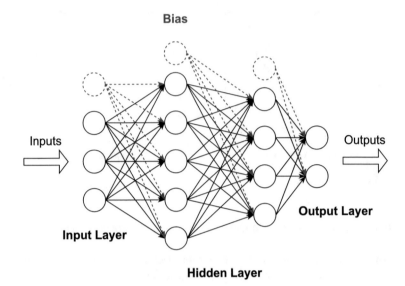

Fig. 1.5 An illustrative example of MLP

$$y^l = a(y^{l-1} W_{l-1,l} + \mathbf{b}_l) \tag{1.6}$$

where y^l is the output of the l-th layer, $W_{l-1,l}$ are the weights between layer $l-1$ and layer l, and $\mathbf{b}_l$ is the bias vector of the lth layer. Note that y^0 is in fact the input feature vector when $l = 1$. Both weights and biases are required to be initialized with random numbers (normally small real numbers from -1 to 1). Common initialization methods include Gaussian initialization, Xavier initialization, and Kaiming initialization [35], among others.

1.1.2.4 Input Layer

The input layer is actually the (preprocessed) input data where each neuron represents one data attribute. And table-format data like credit card [36] and bank marketing [37] are intrinsically well suited to MLP neural networks. Furthermore, other types of data can also be processed and converted to fit the structure of MLPs. For instance, image data can be easily transformed into the input vector whose elements are just pixel values, and time-series data can be partitioned along the sequence direction to several input vectors. However, the MLP neural network, for instance, is not able to extract spatial and time sequence information for image data and time-series data, respectively.

1.1.2.5 Hidden Layer

Layers after the input layer except the last layer are hidden layers. The width of MLPs means the number of neurons of hidden layers and the depth represents the number of hidden layers. And deep neural networks often refer to those networks having many hidden layers. By contrast, wide neural networks represent a network with a large number of neurons per layer.

1.1.2.6 Output Layer

The last layer of MLPs is called the output layer that outputs a scalar or a vector corresponding to the requirement of the learning task. And both the activation function and the number of neurons for the output layer is constrained by the modeling type.

For a regression problem, the output layer may contain only one neuron with no activation function (which can be seen as a linear activation function). Similarly, for a simple binary classification problem, the output layer may also have one neuron using the sigmoid function to output a value between 0 and 1, representing the probability of predicting class 1. In addition, a multi-class classification problem may have multiple neurons with softmax activation function in the output layer and each neuron outputs the probability of predicting one class value.

1.1.2.7 Loss Function

For a typical supervised learning task, the loss function $\ell(\hat{y}_{out}, y_{out})$ representing the distance between the desired label y_{out} and the real prediction $\hat{y}_{out}$ should be constructed before parameter optimization. Two most commonly used loss function for both classification and regression problems will be introduced below.

The first widely used loss function is the cross entropy loss [38] for multi-class classification problems as shown in Eq. (1.7):

$$\ell(\hat{y}_{out}, y_{out}) = -\sum_{c=1}^{M} y_{out,c}\log(\hat{y}_{out,c}) \tag{1.7}$$

where c is the class index and M is the total number of classes. It should be emphasized that the data label y_{out} requires to be converted into a M-length binary one-hot code [39] where only the corresponding position of label class is filled with 1. For instance, for a 4-class classification problem, the label 2 is converted into a binary code 0100 as shown in Fig. 1.6. Furthermore, the prediction value $\hat{y}_{out,c}$ of each class c follows a discrete probability distribution satisfying $\sum_{c=1}^{M} \hat{y}_{out,c} = 1$.

Intuitively, the cross entropy loss aims to make two vectors $\hat{y}_{out}$ and y_{out} become similar. Since y_{out} is one-hot encoded, it is easy to find that the formula

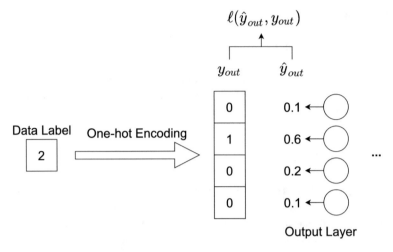

Fig. 1.6 An example of cross entropy loss

$y_{out,c}\log(\hat{y}_{out,c})$ is 0 if c is not equal to the label class. Therefore, the original cross entropy loss can be simplified into Eq. (1.8):

$$\begin{aligned} \ell(\hat{y}_{out}, y_{out}) &= -y_{out,c'}\log(\hat{y}_{out,c'}) \\ &= -\log(\hat{y}_{out,c'}) \end{aligned} \tag{1.8}$$

where c' represents the label class number. Considering that log function is monotonically increasing (monotonically decreasing for negative log function) and $\hat{y}_{out,c'}$ ranges from 0 to 1, minimizing $\ell(\hat{y}_{out}, y_{out})$ is in fact letting $\hat{y}_{out,c'}$ approach to the true label element $y_{out,c'} = 1$.

Besides, the cross entropy loss function can be modified into a binary cross entropy loss to fit the scope of binary classification problems as shown in Eq. (1.9). In this case, the output layer contains one neuron only and y_{out} is a binary number with a value of either 0 or 1.

$$\ell(\hat{y}_{out}, y_{out}) = -y_{out}\log(\hat{y}_{out}) - (1 - y_{out})\log(1 - \hat{y}_{out}) \tag{1.9}$$

The other commonly used loss function for regression problems is the squared error shown in Eq. (1.10):

$$\ell(\hat{y}_{out}, y_{out}) = \frac{1}{2}(y_{out} - \hat{y}_{out})^2 \tag{1.10}$$

where both $\hat{y}_{out}$ and y_{out} are real-valued scalars. Since Eq. (1.10) is a quadratic function achieving the minimum loss value if $\hat{y}_{out} = y_{out}$, thus, minimizing the squared error function will make the prediction $\hat{y}_{out}$ close to the actual label y_{out}.

The above introduced loss functions are constructed by a single data entry. In most scenarios, however, multiple training data are required to be simultaneously imported into the MLP neural network. And the averaged loss function is often adopted to deal with this situation. For convenience, θ is used here to represent both weights and biases, and the averaged loss function for N data samples can be reformulated as Eq. (1.11):

$$L(\theta, \mathbf{X}) = \frac{1}{N} \sum_i \ell(\theta, \mathbf{x}_i) \quad \mathbf{x}_i \in \{\mathbf{x}_1, \mathbf{x}_2 \ldots, \mathbf{x}_N\} \tag{1.11}$$

where $\mathbf{x}_i$ is the ith training data vector and N is the data size. The objective of model training is to find suitable model parameters θ to minimize the expected loss of N data entries.

1.1.2.8 Gradient-Based Optimization Methods

The gradient based parameter optimization method [40] is the most popular MLP training algorithm used during back-propagation due to its efficiency and fast convergence.

Given that the total training data size is N, the batch size is B and the entire data are evenly divided into $\frac{N}{B}$ mini-batches, a typical mini-batch gradient descent (GD) algorithm in each iteration is performed in Eq. (1.12):

$$g_t = \frac{1}{B} \nabla_\theta L(\theta, \mathbf{x}_{t:t+B})$$
$$\theta_{t+1} = \theta_t - \eta g_t \tag{1.12}$$

where g_t is the expected gradients of a B-size mini-batch data at the t-th iteration and η is the learning rate controlling the training footsteps. In each iteration of the MLP training, the averaged model gradients of a randomly selected (without replacement) min-batch data are computed layer by layer. After that, the model parameters θ_{t+1} for the next iteration can be updated by subtracting the product ηg_t.

Note that if the batch size $B = 1$, Eq. (1.12) becomes the standard stochastic gradient descent (SGD) in which the model parameters θ are immediately upgraded once the gradients for a random data sample are computed. However, SGD may cause instability in training. By contrast, the standard GD is another extreme compared to SGD, where the batch size B is equal to the entire data size N, and thus the average gradients for all N training are calculated and subtracted at each iteration. The weakness of the GD is that the computational consumption in each iteration is intensive when the training data is huge and it is more likely to be trapped into the local minimum.

Consequently, to strike a good balance between SGD and GD, mini-batch SGD is usually selected as the training algorithm by setting $1 < B < N$. And nowadays, the usage of these terms are not very strict, and 'GD', 'SGD' and 'mini-batch SGD' all refer to the batched GD learning method. More detailed descriptions of the gradient based methods are presented in Sect. 1.1.6.

1.1.2.9   Overfitting and Dropout

Overfitting is a very common phenomenon occurred in training MLP models. It means that an MLP model learns 'too well' against the training data, making it fail to fit unseen data. As a simple regression example shown in Fig. 1.7, the noisy data samples are fitted to the true function (orange line) and a regression model (blue line). Although the blue line seems to make a perfect fit of the training data, it generalize unseen data poorly.

In general, a machine learning model containing more trainable (effective) parameters is more likely to overfit [41], especially for the cases when the training data are limited. And dropout [42, 43] is a widely used regularization technique to mitigate the overfitting problem for MLPs. As shown in Fig. 1.8(a), dropout is implemented by randomly dropping some neurons during the training. Thus, only parts of model parameters of the MLP network are updated in each iteration.

To understand why dropout is effective, we can consider an extreme situation in which each hidden layer only contains one neuron as shown in Fig. 1.8(b). For this scenarios, the number of model parameters for the slim MLP is too small to remember enough *detailed information* of the training data, effectively avoiding learning 'too well'.

1.1.2.10   Summary

Overall, the MLP neural network is a classic feed-forward artificial neural network. MLPs are capable of learning the representation from the training data and mapping them to the desired prediction outputs. And for a typical supervised learning problem, model training is often equivalent to minimizing the loss between the model predictions and the corresponding data labels. Moreover, regularization techniques

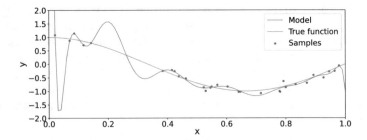

Fig. 1.7 An example of overfitting for a regression problem, where the blue line is the prediction of the model, and the orange line is the true function

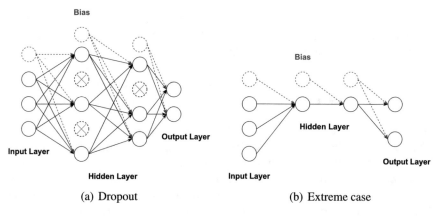

(a) Dropout   (b) Extreme case

Fig. 1.8 An example of dropout and one extreme case in which only one hidden neuron is kept

like dropout are usually adopted to alleviate the overfitting issues caused by limited training data.

However, the MLP neural network is not good at extracting spatial and sequence information of the training data and has gradually replaced by the convolutional neural network and recurrent neural network, respectively, in the field of image classification and speech recognition. Given sufficient training data, it is possible to build a deeper and deeper structure of neural networks without considering too much about the overfitting problems in modern machine learning tasks. In the next two sections, the convolutional neural network and the long short term memory will be introduced.

1.1.3 Convolutional Neural Networks

In the previous section, we focus on learning tabular data samples using MLPs, in which each sample consists of rows (samples) and columns (features), respectively. In this way, we can feed this type of data into models with a structure similar to multi-layer perceptrons. For an image input, we can simply neglect its spatial structure and view it as an one-dimensional tensor. Unfortunately, the unfolded tensor cannot reflect the interactions or relationships between features.

One may ask if one can use image data with inner feature connections as the input of a neural network. The answer is yes. *Convolutional neural networks* (CNNs), a classic and powerful family of deep learning models that are designed for computer vision tasks and have achieved incredible progresses in many fields. CNNs are usually formed by the convolution layers, pooling layers, dropout layers, batch normalization layers and fully connected layers.

In this section, we will give a brief introduction to convolutional neural networks, including the fundamental theories of convolution, the applications of convolution in image processing, the effects of padding, stride, and pooling.

1.1.3.1 Convolution

First of all, consider an one-dimensional convolution operation in signal processing, where a time-varying signal function $f(t)$, and a response function $g(t)$ corresponding to unit step input δ are given, as shown in Fig. 1.9.

It is clear that, at time $t = \tau$, the influence of the signal at time 0 on the system will become $f(0)g(\tau - 0)$, and the influence of the signal at time t is $f(t)g(\tau - t)$. Then, the sum of all input-response pairs in the time window 0 to τ can be calculated by

$$(f * g)(\tau) = \sum_{t=-\infty}^{\infty} f(t)g(\tau - t) \tag{1.13}$$

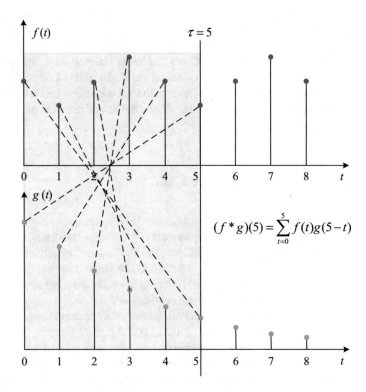

Fig. 1.9 An illustration of convolution

where $(f * g)(\tau)$ is what we call convolution. This equation shows that convolution can be regarded as the overlap of f and g when one of them is flipped and shifted by τ. For continuous time step t, the convolution operation can be defined as

$$(f * g)(\tau) = \int f(t)g(\tau - t)dt. \tag{1.14}$$

1.1.3.2 From Mean Filter to Image Convolution

The calculation of 1D convolution is not difficult and easy to understand, but how does convolution work in 2D cases? In the following, we will use the mean filter to help understand this process.

The mean filter is a typical and prevalent sliding-window linear spatial feature filter that replaces the pixel value with the average value of its neighbors, including itself. Given a gray image, we can calculate the 3×3 mean filter of an image by the following steps:

- Step 1: For a pixel that is not on the edge, get the sum of its 8 neighborhoods and itself. And for the edge pixels, the missing neighborhoods are replaced by 0.
- Step 2: Replace the pixel values with one-ninth of the sum.

The mean filter is easy to implement and is often used to reduce the noise in an image, since the averaging operation is able to eliminate pixel values which are not consistent with their neighbors, as shown in Fig. 1.10. Moreover, we can adopt integral image (also known as the summed area table, SAT) [44] to reduce the computational complexity.

Interestingly, we can transfer the mean filter to the Hadamard product of the original image and a weight matrix with all $\frac{1}{9}$ values, as shown in Eq. (1.15),

$$\begin{bmatrix} * & * & * \\ * & 4 & * \\ * & * & * \end{bmatrix} = \sum \left(\begin{bmatrix} 6 & 7 & 3 \\ 4 & 2 & 4 \\ 5 & 3 & 2 \end{bmatrix} \odot \begin{bmatrix} \frac{1}{9} & \frac{1}{9} & \frac{1}{9} \\ \frac{1}{9} & \frac{1}{9} & \frac{1}{9} \\ \frac{1}{9} & \frac{1}{9} & \frac{1}{9} \end{bmatrix} \right). \tag{1.15}$$

Intuitively, different pixel values will be obtained if we vary the values in the weight matrix, which means that different features are extracted. And it is possible

Original image   **Image with salt noise**   **Denoised image via mean filter**

Fig. 1.10 An example of the mean filter in image processing. The image in the middle has salt and pepper noise added, and the image on the right is processed by the mean filter

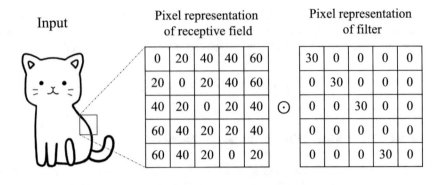

Fig. 1.11 An illustration of image convolution

to find a weight matrix that is sensitive to a certain type of feature, such as rectangular, circle or line. Then one question is, given a feature, can we find a weight matrix that is sensitive to it? The answer is yes, and this is one of the motivations of convolution for images, as shown in Fig. 1.11.

1.1.3.3 Input Layer

Routinely, the raw or processed data samples can be regarded as the input layer of a convolutional neural network, including but not limited to videos, colorful/gray 2D images, and even 1D arrays. For pre-processing, a variety of data augmentation strategies, such as cutout, Gaussian blur and flip, can be employed to enhance the feature extraction capability of the model.

1.1.3.4 Padding and Stride

When using kernel/filter, the pixel values that lie on the edges usually will not be averaged since the lack of neighbors. To avoid this, one straightforward solution is to add extra pixels of filler around the boundary of the input image, namely padding, as shown in Fig. 1.12.

In order to scan the entire input without a full size kernel, we start with the kernel from the top left to the bottom right when computing the correlations between the input image and the sliding window. Intuitively, the number of steps the kernel slides each time is called slides.

Fig. 1.12 An example of zero padding of size 1

0	0	0	0	0	0	0
0	0	20	40	40	60	0
0	20	0	20	40	60	0
0	40	20	0	20	40	0
0	60	40	20	20	40	0
0	60	40	20	0	20	0
0	0	0	0	0	0	0

1.1.3.5 Convolution Layer

In general, a convolution layer in CNNs is to extract the important features that we want by the optimized kernel matrix. The difference between convolution kernel and the mean filter lies in whether the values of the kernel are fixed or not. Figure 1.13 illustrates a typical convolution layer and its calculation process, in which an input image with 3 channels are processed by two convolution kernel (each with three channels, which is the same as the input channel).

As we know, an image is a matrix, or an array, consisting of pixels arranged in columns, rows and channel(s). Let $x = \{x_1, \ldots, x_{c_i}\} \in \mathbb{R}^{c_i \times h_i \times h_i}$ be the input of a convolution layer, the padding size is p, where c_i, h_i are the number of channels of the image (for example, RGB image has 3 channels) and the input width (suppose the height and width are equal). Assume there are c_o convolution kernels, $\{\Phi_1, \Phi_2, \ldots, \Phi_{c_o}\}$, the convolution movement is performed with a step size s, where the first j convolution kernels $\Phi_j \in \mathbb{R}^{c_i \times f \times f}$, f is the kernel size. Take ReLU as the activation function, then the main operation is to use convolution kernel, or filter, to extract the features:

$$y_j = ReLU\left(\sum_{i=1}^{c_i} x_i \odot \Phi_j + b_j\right) \in \mathbb{R}^{h_o \times h_o}, \text{ for } j \in 1, 2, 3, \ldots, c_o, \quad (1.16)$$

where x_i is the i channel of the input, $\odot$ is the convolution operation, the output width h_o is calculated by

$$h_o = 1 + \frac{h_i - f + 2p}{s}. \quad (1.17)$$

Then the output of the input data after passing through the convolutional layer is $y = \{y_1, y_2, \ldots, y_{c_o}\} \in \mathbb{R}^{c_o \times h_o \times h_o}$. From the Eq. (1.16), it is clear that the channel number of a convolution kernel is equal to the input channel c_i, and the number of convolution kernel is the same as the output channel c_o.

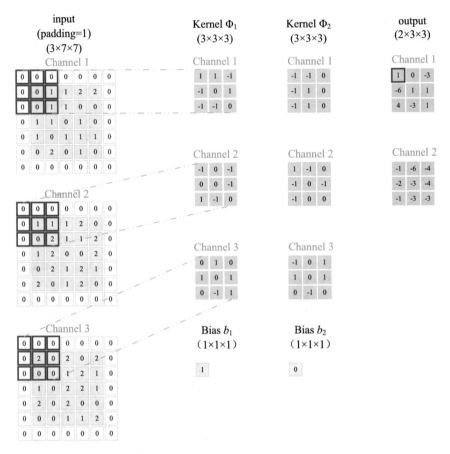

Fig. 1.13 An illustration of the convolution layer

It is not difficult to find that the parameter quantity of a convolution layer ($params$) is determined by the convolution kernels (number of kernel c_o and number of kernel channel c_i), which can be calculated by

$$params = c_o \times (c_i \times f \times f + 1), \tag{1.18}$$

where f is the kernel size (we consider a kernel with the same width and height), the term 1 is the bias for each kernel. If the bias is disabled in each kernel, then the adding 1 should be removed.

Except for $params$, another key indicator of a convolution layer is $FLOPs$, the abbreviation of floating point operations. In a convolution layer, the operations can be categorized into two main steps, multiplication and addition. For multiplication operations, each channel of a kernel will be multiplied with the corresponding input channel, and the amount of element-wise operations is $c_i \times f \times f$. For addition operations, the amount is $(c_i \times f \times f - 1) + 1$, where the last 1 is the bias. Then we have

$$FLOPs = [(c_i \times f \times f) + (c_i \times f \times f - 1) + 1] \times c_o \times h_o \times h_o. \quad (1.19)$$

1.1.3.6 Pooling

Pooling, or downsampling, is one of the key components of convolutional neural network, which is designed and used for the feature selection of the receptive fields to extract the most representative features. It is common to insert a pooling layer between two sequential convolutional layers, which enables dimensionality reduction and parameter reduction, and alleviates overfitting. Maximum Pooling, average pooling and sum pooling are the most commonly used pooling strategies, and the definitions of them are listed as follows:

1. Maximum Pooling: Replace the raw pixel values by the maximum value in the pooling window.
2. Average pooling: Replace the raw pixel values by the average value in the pooling window.
3. Sum pooling: Replace the raw pixel values with the sum of all values in the pooling window.

Generally, 2×2 max pooling operation with a stride of 2 is the most prevalent in the designs of CNNs, since 75% activation can be discarded. Each pooling operation has two parameters, kernel size f and stride s, given an input feature map with size of $h_i \times h_i \times c_i$, the output feature map size $h_o \times h_o \times c_o$ can be deduced by

$$h_o = (h_i - f)/s + 1, \quad (1.20)$$
$$c_o = c_i. \quad (1.21)$$

1.1.3.7 Batch Normalization

Batch normalization (a.k.a. BN), a prevalent method in the context of deep learning has been proven to be effective for overcoming overfitting [32]. As summarized in Algorithm 1, BN layer aims to normalize the activation of a mini-batch by the calculated mean and variance. With batch normalization, the activation among different layers will be pulled to the same distribution $N(0, 1)$, and hence the convergence of the network can be accelerated, and the exploding/vanishing gradient problem can be mitigated.

Algorithm 1 Batch Normalization [32], applied to activation x over a mini-batch.

Input: Values of x over a mini-batch: $\mathcal{B} = \{x_{1...m}\}$; Parameters to be learned: γ, β
Output: $\{y_i = BN_{\gamma,\beta}(x_i)\}$

$$\mu_{\mathcal{B}} \leftarrow \frac{1}{m} \sum_{i=1}^{m} x_i \qquad\qquad\qquad\qquad // \text{ mini-batch mean}$$

$$\sigma_{\mathcal{B}}^2 \leftarrow \frac{1}{m} \sum_{i=1}^{m} (x_i - \mu_{\mathcal{B}})^2 \qquad\qquad\qquad // \text{ mini-batch variance}$$

$$\widehat{x}_i \leftarrow \frac{x_i - \mu_{\mathcal{B}}}{\sqrt{\sigma_{\mathcal{B}}^2 + \epsilon}} \qquad\qquad\qquad\qquad // \text{ normalize}$$

$$y_i \leftarrow \gamma \widehat{x}_i + \beta \equiv BN_{\gamma,\beta}(x_i) \qquad\qquad // \text{ scale and shift}$$

Nevertheless, the BN layer may not work well when the batch size is too small. To address this, group normalization can be adopted as a proxy strategy of BN in the design of networks [33].

1.1.4 Long Short-Term Memory

Unlike previously introduced feed-forward neural networks, long short-term memory (LSTM) [45] is a kind of recurrent neural network with feedback connections. The recurrent neural network (RNN) is designed to process sequence-typed training data and is applicable to domains like machine translation, speech recognition and so on. Before introducing LSTM, the standard RNN would be discussed at first.

1.1.4.1 Recurrent Neural Networks

A simple example of RNN is shown in Fig. 1.14, the outputs of neurons in the recurrent layer (hidden layer) are fully connected to their inputs to become the so called 'recurrent connections'. In this way, if a T-length sequence data are sequentially processed to the RNN step by step, the model prediction $\hat{\mathbf{y}}_t$ of step t is contributed to not only the input $\mathbf{x}_t$ of the current step, but also the output of the hidden layer $\mathbf{h}_{t-1}$ from the last step.

In order to give a clearer description, the model structure of the RNN can also be displayed by a simplified block pattern in Fig. 1.15, where its left panel is a compact scratch of the RNN and the right panel is the unfolded situation. The recurrent layer in Fig. 1.14 is represented by a concise rectangle block *cell* containing the hidden state vector $\mathbf{h}$ (actually the output of the recurrent layer), and all trainable parameters, such as model weights $\mathbf{U}$, $\mathbf{W}$ and $\mathbf{V}$, are shared along all the sequence steps. And each input vector $\mathbf{x}_t$ has its corresponding output vector $\hat{\mathbf{y}}_t$, indicating that $\mathbf{x} = \{\mathbf{x}_1, \mathbf{x}_2, \ldots, \mathbf{x}_T\}$

Fig. 1.14 An illustration of
a simple RNN

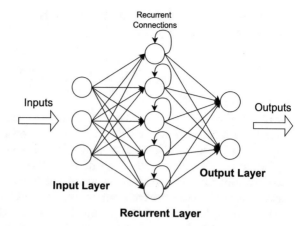

and $\mathbf{y} = \{\mathbf{y}_1, \mathbf{y}_2, \ldots, \mathbf{y}_T\}$ for one data sample have exactly the same sequence length
(many to many). However, this is not always the case in many scenarios and it is
likely that $\mathbf{x}$ and $\mathbf{y}$ are in different lengths. In extreme cases, only one vector $\hat{\mathbf{y}}_T$ of
the last step is outputted for the whole input sequence $\mathbf{x}$ (many to one).

According to Fig. 1.15, the formula of forward propagation for any step t is
performed in Eq. (1.22):

$$\begin{aligned}
\mathbf{h}_t &= a_h(\mathbf{U}\mathbf{x}_t + \mathbf{V}\mathbf{h}_{t-1} + \mathbf{b}_h) \\
\hat{\mathbf{y}}_t &= a_{\hat{y}}(\mathbf{W}\mathbf{h}_t + \mathbf{b}_{\hat{y}})
\end{aligned} \tag{1.22}$$

where $\mathbf{U}$, $\mathbf{W}$ and $\mathbf{V}$ are model weights, $\mathbf{b}$ is the bias and a is the activation function
(can be different for the hidden and output layer). It is clear to find that the hidden
state $\mathbf{h}_t$ of step t is recursively accumulated by all previous $t - 1$ hidden states. And
if there is no special circumstances, the original hidden state $\mathbf{h}_0$ is usually set to be
a zero vector.

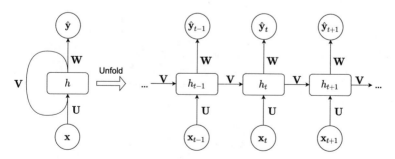

Fig. 1.15 The block figures of folded and unfolded RNNs

The loss of an entire data sequence L can be constructed by *summing* T-step losses shown in Eq. (1.23):

$$L(\hat{\mathbf{y}}, \mathbf{y}) = \sum_{t=1}^{T} \ell(\hat{\mathbf{y}}_t, \mathbf{y}_t) \tag{1.23}$$

where $\mathbf{y}_t$ is the label vector of step t. $\ell(\hat{\mathbf{y}}_t, \mathbf{y}_t)$ is the loss of one single sequence and the loss functions used in feed-forward neural networks, such as cross entropy loss introduced in Sect. 1.1.2.7, are also applicable to RNNs.

The back-propagation of RNNs are different from above mentioned feed-forward networks. Apart from the gradients accumulated layer by layer like the feed-forward network does, the derivatives of recurrent connections for each step t are also propagated inversely through the sequence direction as shown in Fig. 1.16. Consequently, the gradients of the last step T with respect to the weights $\mathbf{U}$ and $\mathbf{V}$ are accumulated by Eq. (1.24), where $\mathbf{U}_1 = \mathbf{U}_2... = \mathbf{U}_T = \mathbf{U}$ and $\mathbf{V}_1 = \mathbf{V}_2... = \mathbf{V}_T = \mathbf{V}$.

$$\begin{aligned}
\frac{\partial \ell_T}{\partial \mathbf{U}} &= \frac{\partial \ell_T}{\partial \mathbf{U}_T} + \frac{\partial \ell_T}{\partial \mathbf{U}_{T-1}} + \cdots + \frac{\partial \ell_T}{\partial \mathbf{U}_1} \\
\frac{\partial \ell_T}{\partial \mathbf{V}} &= \frac{\partial \ell_T}{\partial \mathbf{V}_{T-1}} + \frac{\partial \ell_T}{\partial \mathbf{V}_{T-1}} + \cdots + \frac{\partial \ell_T}{\partial \mathbf{V}_0}
\end{aligned} \tag{1.24}$$

Technically, both $\frac{\partial \ell_T}{\partial \mathbf{U}}$ and $\frac{\partial \ell_T}{\partial \mathbf{V}}$ are summed by the derivatives with respect to $\mathbf{U}_t$ and $\mathbf{V}_t$ of each sequence step t, respectively. In essence, this phenomenon is caused by forward propagation of the hidden state $\mathbf{h}_t$ through the sequence direction (Eq. (1.22)). Let us go through a simple example with a total of three steps ($T = 3$), and the bias $\mathbf{b}_h$ is *removed* for a brevity's sake. And then, the hidden state $\mathbf{h}_3$ is calculated in Eq. (1.25):

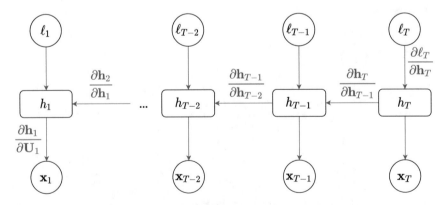

Fig. 1.16 The back-propagation of an unfolded RNN

$$\mathbf{h}_3 = a_h(\mathbf{U}\mathbf{x}_3 + \mathbf{V}\mathbf{h}_2)$$
$$\mathbf{h}_2 = a_h(\mathbf{U}\mathbf{x}_2 + \mathbf{V}\mathbf{h}_1) \tag{1.25}$$
$$\mathbf{h}_1 = a_h(\mathbf{U}\mathbf{x}_1 + \mathbf{V}\mathbf{h}_0)$$

Consequently, the gradients $\frac{\partial \ell_3}{\partial \mathbf{U}}$ can be calculated by the chain rule in Eq. (1.26):

$$
\begin{aligned}
\frac{\partial \ell_3}{\partial \mathbf{U}} &= \frac{\partial \ell_3}{\partial \mathbf{h}_3}\frac{\partial \mathbf{h}_3}{\partial \mathbf{U}} + (\mathbf{V}\mathbf{h}_2)^{'} \\
&= \frac{\partial \ell_3}{\partial \mathbf{h}_3}\frac{\partial \mathbf{h}_3}{\partial \mathbf{U}} + \frac{\partial \ell_3}{\partial \mathbf{h}_3}\frac{\partial \mathbf{h}_3}{\partial \mathbf{h}_2}\frac{\partial \mathbf{h}_2}{\partial \mathbf{U}} + (\mathbf{V}\mathbf{h}_1)^{'} \\
&= \frac{\partial \ell_3}{\partial \mathbf{h}_3}\frac{\partial \mathbf{h}_3}{\partial \mathbf{U}} + \frac{\partial \ell_3}{\partial \mathbf{h}_3}\frac{\partial \mathbf{h}_3}{\partial \mathbf{h}_2}\frac{\partial \mathbf{h}_2}{\partial \mathbf{U}} + \frac{\partial \ell_3}{\partial \mathbf{h}_3}\frac{\partial \mathbf{h}_3}{\partial \mathbf{h}_2}\frac{\partial \mathbf{h}_2}{\partial \mathbf{h}_1}\frac{\partial \mathbf{h}_1}{\partial \mathbf{U}} \\
&= \frac{\partial \ell_3}{\partial \mathbf{U}_3} + \frac{\partial \ell_3}{\partial \mathbf{U}_2} + \frac{\partial \ell_3}{\partial \mathbf{U}_1}
\end{aligned}
\tag{1.26}
$$

Note that, $\frac{\partial \mathbf{h}_t}{\partial \mathbf{U}} = \frac{\partial \mathbf{h}_t}{\partial \mathbf{a}_h}\frac{\partial \mathbf{a}_h}{\partial \mathbf{U}}$ and $\frac{\partial \ell_3}{\partial \mathbf{h}_3} = \frac{\partial \ell_3}{\partial \hat{\mathbf{y}}_3}\frac{\partial \hat{\mathbf{y}}_3}{\partial \mathbf{a}_h}\frac{\partial \mathbf{a}_h}{\partial \mathbf{h}_3}$ which are left out for convenience. Similarly, $\frac{\partial \ell_T}{\partial \mathbf{V}}$ is also computed by both the chain rule and Leibniz rule [46] described in Eq. (1.27):

$$
\begin{aligned}
\frac{\partial \ell_3}{\partial \mathbf{V}} &= \frac{\partial \ell_3}{\partial \mathbf{h}_3}(\mathbf{V}^{'}\mathbf{h}_2 + \mathbf{V}\mathbf{h}_2^{'}) \\
&= \frac{\partial \ell_3}{\partial \mathbf{h}_3}\left(\frac{\partial \mathbf{h}_3}{\partial \mathbf{V}} + \frac{\partial \mathbf{h}_3}{\partial \mathbf{h}_2}\mathbf{h}_2^{'}\right) \\
&= \frac{\partial \ell_3}{\partial \mathbf{h}_3}\frac{\partial \mathbf{h}_3}{\partial \mathbf{V}} + \frac{\partial \ell_3}{\partial \mathbf{h}_3}\frac{\partial \mathbf{h}_3}{\partial \mathbf{h}_2}\left(\frac{\partial \mathbf{h}_2}{\partial \mathbf{V}} + \frac{\partial \mathbf{h}_2}{\partial \mathbf{h}_1}\mathbf{h}_1^{'}\right) \\
&= \frac{\partial \ell_3}{\partial \mathbf{h}_3}\frac{\partial \mathbf{h}_3}{\partial \mathbf{V}} + \frac{\partial \ell_3}{\partial \mathbf{h}_3}\frac{\partial \mathbf{h}_3}{\partial \mathbf{h}_2}\frac{\partial \mathbf{h}_2}{\partial \mathbf{V}} + \frac{\partial \ell_3}{\partial \mathbf{h}_3}\frac{\partial \mathbf{h}_3}{\partial \mathbf{h}_2}\frac{\partial \mathbf{h}_2}{\partial \mathbf{h}_1}\frac{\partial \mathbf{h}_1}{\partial \mathbf{V}} \\
&= \frac{\partial \ell_3}{\partial \mathbf{V}_2} + \frac{\partial \ell_3}{\partial \mathbf{V}_1} + \frac{\partial \ell_3}{\partial \mathbf{V}_0}
\end{aligned}
\tag{1.27}
$$

It is good to see that both Eqs. (1.26) and (1.27) match the formulas in Eq. (1.24). Furthermore, these three equations can be re-formulated into Eq. (1.28):

$$
\begin{aligned}
\frac{\partial \ell_T}{\partial \mathbf{U}} &= \sum_{k=1}^{T}\frac{\partial \ell_T}{\partial \mathbf{h}_T}\left(\prod_{j=k}^{T-1}\frac{\partial \mathbf{h}_{j+1}}{\partial \mathbf{h}_j}\right)\frac{\partial \mathbf{h}_k}{\partial \mathbf{U}} \\
\frac{\partial \ell_T}{\partial \mathbf{V}} &= \sum_{k=1}^{T}\frac{\partial \ell_T}{\partial \mathbf{h}_T}\left(\prod_{j=k}^{T-1}\frac{\partial \mathbf{h}_{j+1}}{\partial \mathbf{h}_j}\right)\frac{\partial \mathbf{h}_k}{\partial \mathbf{V}}
\end{aligned}
\tag{1.28}
$$

where $\prod_{j=k}^{T-1}\frac{\partial \mathbf{h}_{j+1}}{\partial \mathbf{h}_j} = \frac{\partial \mathbf{h}_T}{\partial \mathbf{h}_{T-1}}\frac{\partial \mathbf{h}_{T-1}}{\partial \mathbf{h}_{T-2}}\cdots\frac{\partial \mathbf{h}_{k+1}}{\partial \mathbf{h}_k}$ is the multiplication chain shown in Fig. 1.16. Therefore, the RNN also suffers from the problem of gradient vanishing if $||\frac{\partial \mathbf{h}_{j+1}}{\partial \mathbf{h}_j}||_2 < 1$, making it hard to track the long-term-dependency relationships.

That is, for example, the prediction $\hat{\mathbf{y}}_T$ at the last step is mainly affected by its nearby input sequences such as $\mathbf{x}_T$ and $\mathbf{x}_{T-1}$, while is not easy to be affected by the inputs far away like $\mathbf{x}_1$. No matter whatever the output is, its corresponding impact within a specific location is 'vanished' to be back propagated to forward locations with a long distance. Besides, the gradients of the output connections are not propagated through the time sequence and can be easily calculated according to Eq. (1.29).

$$\frac{\partial \ell_T}{\partial \mathbf{W}} = \frac{\partial \ell_T}{\partial \hat{\mathbf{y}}_T} \frac{\partial \hat{\mathbf{y}}_T}{\partial a_{\hat{y}}} \frac{\partial a_{\hat{y}}}{\partial \mathbf{W}} \tag{1.29}$$

Recall that, the total loss function $L(\hat{\mathbf{y}}, \mathbf{y})$ sums the losses of all T steps. Thus, the gradients with respect to $\mathbf{U}$ and $\mathbf{V}$ over the whole sequence steps are aggregated in Eq. (1.30)

$$\frac{\partial L}{\partial \mathbf{U}} = \sum_{t=1}^{T} \frac{\partial \ell_t}{\partial \mathbf{U}}$$

$$\frac{\partial L}{\partial \mathbf{V}} = \sum_{t=1}^{T} \frac{\partial \ell_t}{\partial \mathbf{V}} \tag{1.30}$$

Therefore, the accumulated gradients for RNNs are often much larger than those for feed-forward neural networks. And the forward and backward pass along the sequence direction make the computational overhead of the RNN very heavy even for the network with only one hidden layer.

1.1.4.2   LSTM Cell

The LSTM neural network is proposed to alleviate the gradient vanishing issues in the standard RNN by *replacing* the RNN cell (rectangle block in Fig. 1.15) with the LSTM cell.

Different from the original RNN cell that directly processes the last state $\mathbf{h}_{t-1}$ to the current state $\mathbf{h}_t$, the LSTM cell additionally introduces a *memory cell* $\mathbf{c}_t$ controlled by three gates. A typical LSTM cell is shown in Fig. 1.17, where σ is the sigmoid function, $\times$ represents element-wise multiplication operation, Γ_f is the forget gate, Γ_u is the update gate, Γ_o is the output gate and $\tilde{\mathbf{c}}_t$ is the replaced memory cell. And then, the forward pass of the LSTM cell can be therefore computed in Eq. 1.31

Fig. 1.17 Forward flow of the LSTM cell

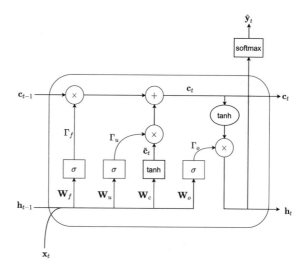

$$\tilde{\mathbf{c}}_t = \tanh(\mathbf{W}_c[\mathbf{h}_{t-1}, \mathbf{x}_t] + \mathbf{b}_c)$$
$$\Gamma_f = \sigma(\mathbf{W}_f[\mathbf{h}_{t-1}, \mathbf{x}_t] + \mathbf{b}_f)$$
$$\Gamma_u = \sigma(\mathbf{W}_u[\mathbf{h}_{t-1}, \mathbf{x}_t] + \mathbf{b}_u)$$
$$\Gamma_o = \sigma(\mathbf{W}_o[\mathbf{h}_{t-1}, \mathbf{x}_t] + \mathbf{b}_o) \qquad (1.31)$$
$$\mathbf{c}_t = \Gamma_u \times \tilde{\mathbf{c}}_t + \Gamma_f \times \mathbf{c}_{t-1}$$
$$\mathbf{h}_t = \Gamma_o \times \tanh(\mathbf{c}_t)$$

Note that, $\mathbf{W}[\mathbf{h}_{t-1}, \mathbf{x}_t]$ is a compact formulation of $\mathbf{W}_{hh}\mathbf{h}_{t-1} + \mathbf{W}_{hx}\mathbf{x}_t$ and $\mathbf{W}$ is equal to $\mathbf{W}_{hh}$ and $\mathbf{W}_{hx}$ concatenated along the column dimension. And σ is the sigmoid function whose output value is close to either 0 or 1 for most input values. Therefore, each element value of Γ is also close to either 0 or 1, approximated to a binary number representing close or open operation of a 'gate'. For a simple example of $\mathbf{c}_t = \Gamma_u \times \tilde{\mathbf{c}}_t + \Gamma_f \times \mathbf{c}_{t-1}$ in Eq. (1.31), if the update gate is 'open' ($\Gamma_u = 1$) and the forget gate is 'closed' ($\Gamma_f = 0$), the memory state of the current step $\mathbf{c}_t$ will be updated by the current replaced memory value $\tilde{\mathbf{c}}_t$ and *forgets* the previous $\mathbf{c}_{t-1}$. By contrast, if the update gate is 'closed' ($\Gamma_u = 0$) and the forget gate is 'open' ($\Gamma_f = 1$), the memory state of the previous step $\mathbf{c}_{t-1}$ will be kept without update.

This implicitly describes the advantage of the LSTM that the memory state is controlled by two gates Γ_u and Γ_f, and $\mathbf{c}_t$ can keep constant until the upgrade requirement is needed. Even for the training data with large sequence length T, it is still possible in the extreme case that the memory state $\mathbf{c}_T = \mathbf{c}_1$. Thus, $\mathbf{c}_1$ in the first step can easily affect the prediction $\mathbf{c}_T$ in the final step, effectively preventing gradient vanishing along the sequence direction. Note that Γ is in fact a vector containing approximated binary numbers other than a single scalar. And in the real-world applications, of course, the gate Γ is not exactly equal to 0 or 1, however, it is really convenient and understandable to regard Γ as a binary vector.

For back-propagation, LSTM works in a similar way to the standard RNN and the intermediate derivatives should be accumulated along both layer and sequence directions. According to Eq. (1.31), four trainable model weights $\mathbf{W}_c$, $\mathbf{W}_f$, $\mathbf{W}_u$ and $\mathbf{W}_o$ should be updated (ignoring the biases for simplicity), the number of which is twice as the standard RNN. The detailed process of back-propagation for the LSTM will not be introduced here.

Overall, the LSTM is a powerful and flexible recurrent neural network proposed to deal with gradient vanishing issues, and the core idea of the LSTM cell is to additionally build a cell memory controlled by the gates to remember or forget the forward state. However, the LSTM neural network consumes more computational resources than the standard RNN due to more trainable parameters and more complex cell structure.

1.1.5 Decision Trees

Although neural networks are powerful and 'solution-to-anything' of machine learning models, they are usually troubled by high computational costs, long training time and weak interpretation. And modern deep learning technologies require a large number of balanced training data and carefully designed structure or hyperparameters to train a qualified model. Therefore, the decision tree can be instead adopted to deal with these issues.

Unlike deep learning models that contain abundant model parameters, the decision tree is a *non-parametric* machine learning model without any trainable parameters, learning the input data by constructing tree nodes with decision rules. Tree-based models routinely outperform neural networks and XGBoost [47] is so far the most popular machine learning algorithm for top-performing competitions. Of course, in the areas of image classifications and text recognition, deep neural networks are still predominant, the decision tree has more advantage in other common uses due to its fast training speed and good interpretation.

1.1.5.1 Model Structure

A typical decision tree has a flowchart-like structure consisting of several tree nodes, each of which contains a decision rule to decide the following path from two branches (can be more) connected to the child node of the next level. It can be used for both classification tasks as well as regression tasks, and the term 'CART' refers to Classification and Regression Tree. Let us consider a simple example of the binary classification decision tree shown in Fig. 1.18, where H is the height, W is the weight, LH represents long hair, S indicates smoking, F is female and M is male.

The node with decision rule $H < 170$ located at the top of the entire tree is defined as the *root* node, the intermediate (parent) node can be further divided into different child nodes is the *decision* node, the bottom node without child node is the *leaf* node

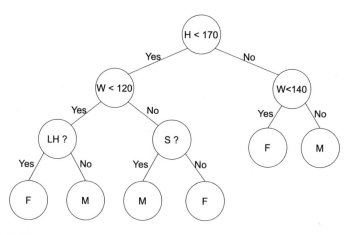

Fig. 1.18 A simple example of the binary classification tree

and the straight line connecting any *parent* node and *child* node is the branch. Note that the outputs in leaf nodes (*F* and *M* in Fig. 1.18) represent leaf weights or leaf values other than the decision rules. If there is a person with 160cm height, 110kg weight and long hair (just an example which may be not reasonable), the person will be easily classified as female according to the corresponding three decision rules.

1.1.5.2 Tree Construction

Training decision trees are totally different from training parametric models like neural networks introduced previously. In fact, the decision tree is already well trained after creating the entire tree. Consequently, 'constructing' or 'building' a decision tree is often adopted for an apt description of model training. The core algorithms of building trees are dependent on the types of leaf values, varied for classification and regression problems.

The main purpose of tree constructions is to find the best splits relying on tree nodes to optimally partition the training data with appropriate decision rules. At the beginning of training, the whole training data are allocated to the root node. And the data splits can be recursively performed on data features according to the predefined split technique.

If a specific data feature containing just discrete numbers, each unique discrete value can be regarded as the split threshold of that feature. Although this method may find the potential optimal split, it must cause unexpected computational costs if the feature range is large and is inapplicable to data features with continuous values. A common solution is to sort and evenly partition the data into, for instance, $q = 3$ buckets as shown in Fig. 1.19, where the original threshold value 6 is reduced to 3, significantly decreasing the computational overhead. Furthermore, this bucketing approach is also well suited for continuous values.

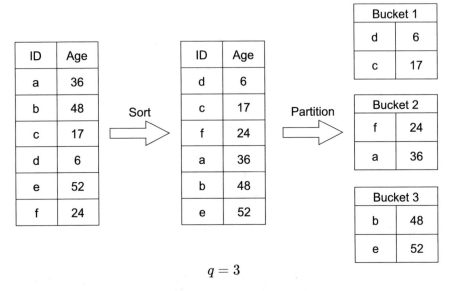

$$q = 3$$

Fig. 1.19 A simple example of split buckets when $q = 3$

For data points of the current tree node, a quantitative 'score' L_{split} of each threshold value within each data feature should be computed, and the split with the maximum score is often regarded as the best split. Two most commonly used split functions will be introduced in the following.

1.1.5.3   Gini Impurity

The term 'purity' means that all the training data are ideally partitioned into correct classes, and Gini impurity is actually a function that gauges the probability of incorrectly classifying a data entry. The mathematical notation of the Gini impurity is given by Eq. (1.32):

$$Gini = 1 - \sum_{i=1}^{n}(p_i)^2 \tag{1.32}$$

where p_i is the probability of picking a data point belonging to class i, with subject to $\sum_{i=1}^{n} p_i = 1$. It is clear to see that Gini impurity ranges from 0 to 1, of which 0 means that all the branch data belong to one certain class with 'pure' division, 1 denotes the branch data are randomly distributed over multiple classes, and 0.5 represents that the branch data points are evenly distributed into several classes.

And then, the quality of the split named *Gini gain* is calculated by subtracting weighted sum of Gini impurity for each branch from Gini impurity of the current parent node p in Eq. (1.33):

$$Gini\ gain = Gini_p - \sum_{j=1}^{C} \frac{n_j}{n} Gini_j \qquad (1.33)$$

where C is the total number of branches, j is the branch index, n is the total data size at the parent node p and $\frac{n_j}{n}$ is the weight of each branch satisfying $\sum_{j=1}^{C} \frac{n_j}{n} = 1$. The higher Gini gain is, the better quality of the split will be, and the best split is selected with the corresponding highest Gini gain.

1.1.5.4 Information Gain

Intuitively speaking, information gain indicates the importance of the data attribute calculated by entropy H. In contrast to purity, entropy describes the degree of disorganization of the branch data, and larger entropy denotes higher degree of disorganization. Entropy and information gain are often combined together to collaboratively build a decision tree depending on the learning algorithm such as the classic iterative dichotomiser 3 (ID3) [48].

At first, entropy $H(S)$ measuring the amount of uncertainty for data sets S is given by Eq. 1.34:

$$H(S) = \sum_{i=1}^{C} -p_i log_2 p_i \qquad (1.34)$$

where S is the dataset to be split, i is the class index, C is the total number of classes, and p_i is the proportion of the number of data points for class i to the total data size in S. When entropy $H(S) = 0$, the dataset S is perfectly classified into the same class.

Similar to what we did in Gini impurity, the final entropy of a split for any attribute A is computed by weighted summing entropy of all branches shown in Eq. (1.35):

$$H(S|A) = \sum_{t \in T} \frac{n_t}{n_S} H(t) \qquad (1.35)$$

where T is a set of split subsets for attribute A satisfying $S = \bigcup_{t \in T} t$ and $\frac{n_t}{n_S}$ is the proportion of the number of samples in subset T to the number of samples in S. And then, information gain $IG(A)$ describing the distance in entropy from before to after is given by the following Eq. (1.36), where higher information gain denotes the split with better quality.

$$IG(S, A) = H(S) - H(S|A) \qquad (1.36)$$

Fig. 1.20 Imperfect split

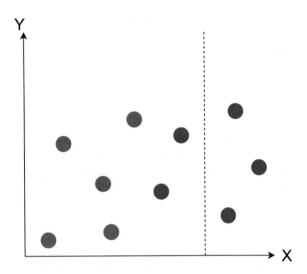

Given a simple example of an imperfect split shown in Fig. 1.20, where each circle dot represents one data point in S. Entropy before the split (dashed line) is $H(s) = -(0.5\log_2 0.5 + 0.5\log_2 0.5) = 1$ according to Eq. (1.34). And similarly, entropy of both left panel and right panel are calculated as $H(\text{left}) = -(\frac{5}{7}\log_2\frac{5}{7} + \frac{2}{7}\log_2\frac{2}{7}) = 0.598$ and $H(\text{right}) = -(0\log_2 0 + \log_2 1) = 0$. Consequently, the information gain is $IG = 1 - (\frac{7}{10} \cdot 0.598 + \frac{3}{10} \cdot 0) = 0.5814$.

In each iterative training of ID3, all the information gains of the remaining attributes are required to be calculated and the attribute with the largest gain will be regarded as the best split. This process is repeated until all the training data are classified.

1.1.5.5   Pruning

Decision trees also suffer from overfitting problems especially for those with many branches and layers. And pruning is one of the most commonly used techniques to prevent overfitting the training data by removing redundant subtrees. Decision tree pruning is often categorized into two types: one is pre-pruning and the other is post-pruning.

Pre-pruning is performed by setting some early-stop rules during the period of tree constructions, and once the condition is satisfied, the current tree node will immediately stop splitting and become the leaf node. The stop condition is usually set as, for example, the information gain is above predefined threshold or the already constructed tree reaches the maximum depth. Note that the tree depth is always evaluated excluding the root node, and thus, the tree depth in Fig. 1.18 is three other than four. Consequently, pre-pruning has an advantage of generating more light weighted tree structure, avoiding the potential risk of data overfitting.

Post-pruning, on the other hand, prunes the tree from bottom to up after the entire model is built. It starts from decision nodes at the bottom and decides whether they will be kept or not based on the measurements of pruning condition like Gini impurity or information gain. Since the entire tree is built only depending on the learning algorithm itself, post-pruning cannot improve efficiency during training.

1.1.6 Gradient-Based Methods

Once the model structure, hyper-parameter configuration and training data are given, an appropriate optimizer for model parameter optimization and loss/objective minimization is necessary. In general, the performance of an optimizer is quite important for machine and deep learning, since the training procedure takes hours, days or even weeks for deep models and large datasets. Here, we will give a brief introduction to the widely adopted gradient based optimization algorithms.

1.1.6.1 Gradient Descent

Gradient descent, a.k.a. GD, is one of the most prevalent optimization algorithm in the early stage of machine learning, although it is rarely used directly in deep learning. However, GD can be viewed as an excellent example for how gradients can help reduce the value of loss.

Let w, b be the weight and bias of a given network to be optimized, and the training dataset $\mathcal{D}$ contains n sample pairs $\{x_i, y_i\}_{i=1}^n$. Then, the logit z can be regarded as

$$z = wx_i + b, \tag{1.37}$$

and the corresponding prediction of the model is

$$\hat{y}_i = \delta(wx_i + b). \tag{1.38}$$

where δ is the activation function, here we take sigmoid function as an example,

$$\delta(z) = \frac{1}{1 + e^{-z}}. \tag{1.39}$$

If we use mean square error as the loss function, for a certain sample (x_i, y_i), its loss l_i is

$$l_i = \frac{1}{2}(y_i - \hat{y}_i)^2, \tag{1.40}$$

where the coefficient $1/2$ is for the convenience of derivation.

Then, for gradient descent optimization method, the optimization objective should be inferred from the mean of all n samples, which is

$$l = \frac{1}{n}\sum_{i=1}^{n} l_i = \frac{1}{2n}\sum_{i=1}^{n}(y_i - \hat{y}_i)^2. \tag{1.41}$$

Therefore, the derivative of l with respect to $\hat{y}$ is

$$\frac{\partial l}{\partial \hat{y}} = \frac{1}{n}\sum_{i=1}^{n} \frac{\partial l_i}{\partial \hat{y}_i} = \frac{1}{n}\sum_{i=1}^{n}(y_i - \hat{y}_i), \tag{1.42}$$

and the derivative of $\hat{y}_i$ with regard to z_i (Sigmoid activation) is

$$\frac{\partial \hat{y}_i}{\partial z_i} = \sigma(z_i)(1 - \sigma(z_i)) = \hat{y}_i(1 - \hat{y}_i). \tag{1.43}$$

In addition, the derivative of z_i regarding x_i is

$$\frac{\partial z_i}{\partial w} = x_i. \tag{1.44}$$

Finally, the gradient of w can be calculated by the chain rule

$$\nabla l(w) = \frac{\partial l}{\partial w} = \frac{1}{n}\sum_{i=1}^{n} \frac{\partial l_i}{\partial \hat{y}_i} \frac{\partial \hat{y}_i}{\partial z_i} \frac{\partial z_i}{\partial w} = \frac{1}{n}\sum_{i=1}^{n}(y_i - \hat{y}_i) \cdot \hat{y}_i \cdot (1 - \hat{y}_i) \cdot x_i. \tag{1.45}$$

Similarly, the bias' gradient is

$$\nabla l(b) = \frac{\partial l}{\partial b} = \frac{1}{n}\sum_{i=1}^{n} \frac{\partial l_i}{\partial \hat{y}_i} \frac{\partial \hat{y}_i}{\partial z_i} \frac{\partial z_i}{\partial b} = \frac{1}{n}\sum_{i=1}^{n}(y_i - \hat{y}_i) \cdot \hat{y}_i \cdot (1 - \hat{y}_i). \tag{1.46}$$

Once the gradients of w and b are obtained, we can use

$$w = w - \eta \nabla l(w), \tag{1.47}$$
$$b = b - \eta \nabla l(b), \tag{1.48}$$

to iterate w and b, where η is usually called learning rate. In this way, the value of l may be declined towards to 0. When using different loss functions and activation functions, for example, cross entropy loss and ReLU activation, the backpropagation process is similar and will not be discussed here.

1.1.6.2 Stochastic Gradient Descent

Intuitively, the computational cost of gradient descent is $O(nd)$, where d is the dimensionality of each sample due to the calculation of the mean of all training samples. Therefore, when the size of training dataset is very large, the total computational cost will be extremely expensive.

To reduce the computational cost of GD without affecting its performance, researchers try to estimate the model gradients in an unbiased way. Stochastic gradient descent (SGD) is one of the most popular solutions, which takes $O(d)$ for each iteration.

At each iteration of stochastic gradient descent, we uniformly and randomly sample a sample pair (x_i, y_i), $i \in \{1, 2, \ldots, n\}$ from all n samples and compute its gradients $\nabla l_i(w)$ and $\nabla l_i(b)$ to update w and b,

$$w = w - \eta \nabla l_i(w), \tag{1.49}$$

$$b = b - \eta \nabla l_i(b), \tag{1.50}$$

where $\nabla l_i(w) = (y_i - \hat{y}_i) \cdot \hat{y}_i \cdot (1 - \hat{y}_i) \cdot x_i$ and $\nabla l_i(b) = (y_i - \hat{y}_i) \cdot \hat{y}_i \cdot (1 - \hat{y}_i)$. The reason why we can use an uniformly selected sample instead of all n samples is that $\nabla l_i(\cdot)$ is an unbiased estimation of $\nabla l(\cdot)$:

$$\mathbb{E}[\nabla l_i(\cdot)] = \frac{1}{n} \sum_{i=1}^{n} \nabla l_i(\cdot) = \nabla l(\cdot). \tag{1.51}$$

Hence, the stochastic gradient is a good estimation of the gradient. Meanwhile, GD has trouble escaping from saddle points, but SGD can overcome this issue because of the introduced noise via the randomly selected samples.

1.1.6.3 Mini-Batch Stochastic Gradient Descent

Although stochastic gradient descent decreases the computational cost of gradient descent from $O(nd)$ to $O(d)$, it may get trapped by the local optimum since only one sample is used in each iteration. Thus, we usually make a compromise between GD and SGD in practice, i.e. mini-batch stochastic gradient descent.

The idea of mini-batch SGD is very simple: uniformly and randomly sample $B(\geq 1)$ samples to compute the gradient in each iteration,

$$\nabla l_i(w) = \frac{1}{B} \sum_{i=1}^{B} (y_i - \hat{y}_i) \cdot \hat{y}_i \cdot (1 - \hat{y}_i) \cdot x_i, \tag{1.52}$$

$$\nabla l_i(w) = \frac{1}{B} \sum_{i=1}^{B} (y_i - \hat{y}_i) \cdot \hat{y}_i \cdot (1 - \hat{y}_i). \tag{1.53}$$

where B is usually called batch size, and mini-batch SGD will be equal to SGD or GD when B varies from 1 to n. Since B samples are uniformly sampled at random from the training dataset, the expectations of the above gradients are still unbiased, but their standard deviation is reduced by a factor of $B^{-\frac{1}{2}}$. Therefore, the mini-batch SGD combines the advantages of GD and SGD, and its computational cost is $O(Bd)$ for a single iteration.

1.1.6.4 Momentum

Mini-batch SGD is an excellent method to accelerate gradient computation, but it will be handicapped when dealing with some tricky ill-conditioned loss functions. For example, $l = 0.01w_1^2 + 5w_2^2$, the derivatives of w_1 and w_2 are $0.02w_1$ and $10w_2$, respectively. In this case, the gradient of w_2 is much higher than w_1 under the same learning rate, which will cause w_2 changes much more rapidly during optimization and we can hardly find a suitable learning rate to make w_1 and w_2 converge at the same time.

For the abovementioned ill-conditioned problem in gradient descent, momentum method is a good solution, in which a *momentum* parameter v is introduced,

$$v_t = \beta_1 v_{t-1} + g_t, \tag{1.54}$$

where t is the index of an iteration, $\beta \in (0, 1)$ is a parameter controlling the magnitude and empirically set to 0.9, g_t is the gradient of a certain parameter (take w as an example).

Then, the model parameter can be updated by

$$w_{t+1} = w_t - \eta v_t. \tag{1.55}$$

As we can see, the implementation of momentum SGD is quite straightforward, but an additional vector need to be stored.

1.1.6.5 AdaGrad

Till now, all gradient descent methods use the first-order gradient for optimization, and fix the learning rate during training. Nevertheless, decreasing the value of learning rate has been proven successful in many tasks.

Hence, adaptive learning rate catches the attention of researchers, and the introduction of the second-order momentum indicates the coming of adaptive learning rate.

AdaGrad is the representative adaptive learning rate algorithm, in which the second-order momentum V_t records the sum of all square gradients,

$$V_t = \sum_{i=1}^{t} g_i^2, \tag{1.56}$$

and then AdaGrad updates the gradient of the parameter (for example, w),

$$w_{t+1} = w_t - \eta \cdot g_t / \sqrt{V_t + \epsilon} \tag{1.57}$$

$$= w_t - \eta / \sqrt{V_t + \epsilon} \cdot g_t, \tag{1.58}$$

it is clear that the learning rate η changes to $\eta / \sqrt{V_t + \epsilon}$, and usually we need to add a tiny constant ϵ to avoid zero denominator. And the lager the second momentum is, the smaller the learning rate will be.

1.1.6.6   RMSProp

AdaGrad records all historical gradients to compute the second momentum for learning rate adaptation, which is too radical in practice. A surrogate solution is to record the gradients in a sliding time window instead of all gradients, namely RMSProp. The modification is quite simple, which can be conducted by:

$$V_t = \beta_2 V_{t-1} + (1 - \beta_2) g_t^2, \tag{1.59}$$

where β_2 is a hyper-parameter determines how long the history is, and hence the update rule of RMSProp is as follows:

$$w_{t+1} = w_t - \eta \cdot g_t / \sqrt{V_t + \epsilon} \tag{1.60}$$

$$= w_t - \eta \cdot g_t / \sqrt{\beta_2 V_{t-1} + (1 - \beta_2) g_t^2} \tag{1.61}$$

Obviously, RMSProp is very similar to AdaGrad, but it uses a time window to adjust the influence of the second-order gradients.

1.1.6.7   Adam

As we can see, the momentum SGD adds the first-order momentum to SGD, RMSProp adds second-order momentum, and when first-order and second-order momentum are combined, the algorithm is known as Adam.

First of all, calculate the first-order momentum at iteration t,

$$v_t = \beta_1 v_{t-1} + g_t. \tag{1.62}$$

Secondly, compute the second-order momentum,

$$V_t = \beta_2 V_{t-1} + (1 - \beta_2) g_t^2. \tag{1.63}$$

In the end, update the parameter by

$$w_{t+1} = w_t - \eta \cdot v_t / \sqrt{V_t + \epsilon} \tag{1.64}$$

$$= w_t - \eta \cdot (\beta_1 v_{t-1} + g_t) / \sqrt{\beta_2 V_{t-1} + (1 - \beta_2) g_t^2 + \epsilon}. \tag{1.65}$$

Adam combines features of many popular optimization algorithms and is quite efficient in many tasks, hence it becomes very popular recently.

1.2 Evolutionary Optimization and Learning

The gradient based methods have been the most popular optimizer for machine learning. These methods are effective, although they require that the loss function (or the objective function) should be continuous and differentiable. In addition, the learning problem is typically assumed to be a convex optimization problem; otherwise, the gradient based methods cannot guarantee to find the global optimum. Finally, gradient based methods may not work properly on discrete or combinatorial optimization problems, e.g., for feature selection or neural architecture search. Therefore, this section presents a class of population-based metaheuristics for optimization, which can be adopted to solve optimization and learning problems. Although there are a large number of different classes of metaheuristics, we introduce only three representative classes, namely genetic algorithms, genetic programming, and particle swarm optimization. We start with a brief introduction to optimization and learning before we describe these evolutionary and swarm optimization algorithms, because they are predominantly used as optimizers in this monograph, although they can also solve problems beyond optimization and learning.

1.2.1 Optimization and Learning

Many real-world problems, including machine learning, can be formulated as an optimization problem. For example, in Sect. 1.1.6, the gradient based methods are used to minimize the loss function, which can be seen a class of optimization problem. More generally, an optimization problem consists of three most important components, i.e., a set of decision variables, an objective function to be minimized or maximized, and a number of constraints to be satisfied. Without the loss of generality, we consider the following constrained minimization problems:

$$\text{minimize} \quad f(\mathbf{x}) \tag{1.66}$$

$$\text{subject to} \quad g_j(\mathbf{x}) < 0, \quad j = 1, 2, \ldots, J. \tag{1.67}$$

$$h_k(\mathbf{x}) = 0, \quad k = 1, 2, \ldots, K. \tag{1.68}$$

$$\mathbf{x}^L \leq \mathbf{x} \leq \mathbf{x}^U. \tag{1.69}$$

In the above, $\mathbf{x}$ represents an n-dimensional decision vector, $f(\mathbf{x})$ is the objective function, $g(\cdot)$ and $h(\cdot)$ are the inequality and equality constraints, respectively. Note that the above constrained problem has J inequality constrains and K equality constraints, and all decision variables are bounded. If both J and K equal 0, and there is no upper or lower bound on the decision variables, this becomes an unconstrained optimization problems. In addition, an optimization problem is called convex, if the objective function is convex and if the feasible region is convex. For a convex optimization problem, every local optimum is a global optimum and it is strictly convex when it has only one optimal solution.

In the above minimization problem, there is only one objective function, which is also known as single-objective optimization problem. In reality, an optimization problem may have multiple objectives to be optimized, which is then called a multi-objective optimization problem. In the following, we will first describe evolutionary and swarm intelligence algorithms for single-objective optimization, and then elaborate on algorithms for multi-objective optimization.

1.2.2 Genetic Algorithms

Genetic algorithms [12] were first proposed by John Holland in the 1970s and extended by David Goldberg in a seminal book [49]. The canonical genetic algorithms were developed for constructing adaptive systems, solving optimization problems, and accomplishing machine learning tasks. Genetic algorithms aimed to simulate natural evolution at different levels, including information encoding at the genetic level, reproduction at the individual level, and survival of the fittest at the population level. Thus, a canonical genetic algorithm works on the basis of a population, which consists of a number of individuals. In the context of optimization, each individual represents a candidate solution, and its objective value is called the fitness of the individual. Similar to biological organisms, each individual is represented by a DNA composed of one or multiple chromosome representing the genotype of the individual. The genotype can then be decoded into its phenotype, representing the values of the decision variables of the candidate solution in the context of optimization. Given the decision variables, the objective of this solution can then be calculated, which is known as the fitness of the individual. Note that for minimization, the smaller the objective value, the better, and the more likely the individual is able to survive.

The map from the genotype to the phenotype plays a pivotal role in a genetic algorithm, depending on which one may distinguish one type of genetic algorithm from another. The most basic genetic algorithm uses either binary or gray coding to represent binary, integer, or even real-valued decision variables. Thus, if the decision variables themselves are not binary, they need to be decoded into integers or real numbers. In the following, we take the hyperparameter optimization of a multi-layer perceptron containing one input layer, one hidden layer and one output layer as an example. In other words, we aim to optimize the number of neurons in the hidden layer and the learning rate of the gradient based method for training the weights. In this case, the number hidden nodes is an integer, and the learning rate is a real-valued number. To represent the number of hidden nodes and the learning rate using binary coding, we usually need to provide a range of the coded values, say the number of hidden nodes is between 0 and 15, and the learning rate is between 0.0 and 0.1. Meanwhile, we assume the number of hidden nodes is encoded using a 4-bit binary string, while the learning rate is encoded by a 3-bit binary string. Thus, if the first string is [0 1 1 0], and the second is [1 0 1], then the number of hidden nodes is decoded to be:

$$0 \times 2^3 + 1 \times 2^2 + 1 \times 2^1 + 0 \times 2^0 \tag{1.70}$$

which equals 6. For the learning rate, we first need to decode the binary string into integer:

$$1 \times 2^2 + 0 \times 2^1 + 1 \times 2^0 \tag{1.71}$$

which results in 5. We can then further convert it into a real-valued learning rate given the lower and upper bounds:

$$0.0 + (0.1 - 0.0)\frac{5}{2^3 - 1} \tag{1.72}$$

which is 0.0714.

More generally, a binary string of l bits can be converted into an integer as below:

$$\sum_{i=1}^{l} 2^{i-1}. \tag{1.73}$$

And it can be further converted into a real number once the lower bound (a) and upper bound (b) are given:

$$a + (b - a)\frac{\sum_{i=1}^{l} 2^{i-1}}{2^{l-1}}. \tag{1.74}$$

Once the representation of the decision variables is specified, a genetic algorithm will start with initializing a population of size N, which needs to be defined beforehand. Usually, N can be specified according to the the dimension of the search space and computational resources available. The larger the search space, a larger population size might be more helpful. Empirically, N ranges from 20 to 100 for medium-size optimization problems containing up to 200 decision variables. These N initial solutions are then evaluated before reproduction and serve as the first parent population. To produce offspring individuals, two parent individuals can be chosen pairwise and then perform a crossover between the two parents to generate two offspring, simulating what happens in biology. For example, given the following two parents, in which the first chromosome (the first four bits) represent the number of hidden nodes, and the second chromosome (last three bits) represent the learning rate:

$$p_1 = [[0110][101]] \tag{1.75}$$
$$p_2 = [[1011][010]] \tag{1.76}$$

Before performing a crossover, we must specify the number of crossover points and their location. The simplest crossover is a 1-point crossover, meaning randomly selecting one location in the genotype and then exchanging the binary bits of the two parents. Assuming we take the position between the third and fourth bits as location to do a 1-point crossover, then the following two offspring will be generated:

$$o_1 = [[0111][010]] \tag{1.77}$$
$$o_2 = [[1010][101]] \tag{1.78}$$

This crossover operation repeats, with a probability, for all individuals in the parent population, generating N offspring individuals, which are new candidate solutions for the problem to be optimized. If the probability is 0, then two parents will be directly passed to the offspring population. Before selection based on the survival of the fittest principle, a mutation operation will be performed on each individual. For binary coded individuals, a mutation operator simply flits each bit at a low probability, i.e., changes '0' to '1', or '1' to '0'.

Once N (the same size as the parent population) offspring individuals are generated, N parent individuals will be selected using a selection mechanism. The most widely used selection mechanism is the fitness proportionate selection, which selects individuals with a probability proportional to their fitness value. That is, the better the individual is, the more likely the individual will be selected. Assume the fitness of the ith individual is f_i, then the probability of the individual being selected as a parent is

$$prob_i = \frac{f_i}{\sum_{i=1}^{N} f_i}. \tag{1.79}$$

Note, however, that for a minimization problem, f_i is not the objective value of the candidate solution. Instead, we need to convert the objective value into a fitness, usually by inverting it:

$$f_i = \frac{1}{f(\mathbf{x}_i) + \epsilon}, \tag{1.80}$$

where $\mathbf{x}_i$ is the candidate solution of the i-th individual, $f(\mathbf{x}_i)$ is its objective value, and $\epsilon > 0$ is a small positive constant to avoid division by zero. A diagram for a

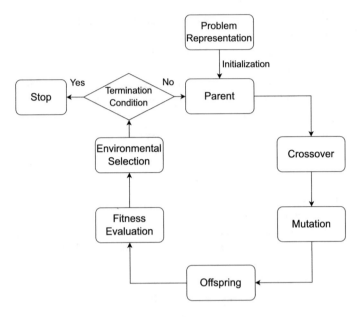

Fig. 1.21 A diagram of a generic genetic algorithm

generic genetic algorithm is given in Fig. 1.21. There are several variants of the canonical genetic algorithm, in which different representations, different crossover, mutation or selection operators can be used. For example, one may apply multi-point crossover or more generally the uniform crossover. In addition to the fitness proportionate selection, rank proportionate or tournament selection can also be adopted. In tournament selection, a subset of the offspring population is randomly chosen, and the best individual of the subset is passed to the next parent population. This repeats N (the population size) times until the next parent population is filled up. The most typical tournament size (the size of the randomly chosen subset) is two and the corresponding tournament selection is known as binary tournament selection.

The most essential change lies in the representation of the genetic algorithm. While the canonical genetic algorithm adopts binary coding for representation of the search space, many variants using real-valued coding have been proposed for continuous optimization, which are known as real-coded genetic algorithms [50]. By real-valued coding, we mean that each decision variable of a continuous optimization problem is directly represented by a real number in the genotype, instead of a binary string. For example, in the previous example, the real-coded chromosome will be [5.0 0.0714], i.e., it codes a solution with five hidden noses and a learning rate of 0.0714.

Given the real-coded representation, the crossover and mutation operators must be changed accordingly. Although several crossover operators are available, the simulated binary crossover [51] and polynomial mutation are the most widely used for real-coded genetic algorithms. Given two real-coded parent individuals

$\mathbf{p}_1 = [p_{11}, p_{12}, \ldots, p_{1n}]$ and $\mathbf{p}_2 = [p_{21}, p_{22}, \ldots, p_{2n}]$, where n is the number of decision variables, then the simulated binary crossover works as follows:

1. Create a random number r between 0 and 1;
2. Calculate a *spread factor* γ as follows:

$$\gamma(r) = \begin{cases} (2r)^{\frac{1}{\eta_c+1}}, & \text{if } r \leq 0.5 \\ \dfrac{1}{2(1-r)^{\frac{1}{\eta_c+1}}}, & \text{if } r > 0.5 \end{cases} \qquad (1.81)$$

where $\eta_c > 1$ is the called the crossover distribution factor. This way, a probability distribution of the spread factor is defined, simulating a contracting binary crossover when $\gamma < 1$ and an expanding crossover when $\gamma > 1$. When $\gamma = 1$, the two offspring individuals are the same as the two parents.

3. The two offspring individuals are calculated as follows:

$$o_{1i} = 0.5[(1 - \gamma)p_{1i} + (1 + \gamma)p_{2i}] \qquad (1.82)$$
$$o_{2i} = 0.5[(1 + \gamma)p_{1i} + (1 - \gamma)p_{2i}] \qquad (1.83)$$

where $i = 1, 2, \ldots, n$.

Similar to the simulated binary crossover, polynomial mutation for real-coded genetic algorithms is also generated based on a specified probability distribution. Assume the i-th decision variables of a real-coded genetic algorithm $x_i \in [a_i, b_i]$, the mutant of x_i, denoted as x_i' can be generated by:

$$x_i' = \begin{cases} x_i + \delta_L(x_i - a_i), & \text{if } u \leq 0.5 \\ x_i + \delta_R(b_i - x_i), & \text{if } u > 0.5 \end{cases} \qquad (1.84)$$

where u is a randomly generated number between $[0, 1]$, δ_L and δ_R are two parameters defined as follows:

$$\delta_L = \quad (2u)^{\frac{1}{1+\eta_m}} - 1, \quad \text{if } u \leq 0.5, \qquad (1.85)$$
$$\delta_R = 1 - (2(1 - u))^{\frac{1}{1+\eta_m}}, \text{if } u > 0.5, \qquad (1.86)$$

where u is a random number between 0 and 1, and η_m is a distribution factor for mutation.

In addition to real-coded genetic algorithms, there are other types of evolutionary algorithms that are efficient for continuous optimization, in particular evolution strategies [13]. Compared to genetic algorithms, one unique feature of evolution strategies is self-adaptation, i.e., some important parameters, such as step sizes of evolution strategies are also encoded in the chromosome and subject to evolution, just like decision variables. The main benefit of self-adaptation is that evolution strategies are able to identify the problem structure and find the shortest path towards the optimum. The most widely used evolution strategy is the covariance adaptation evolution strategy [52], which adapts the step sizes for each decision variables, and the correlation between the decision variables, making it particularly powerful for continuous optimization problems with correlation between the decision variables.

It should be mentioned that genetic algorithms can use hybrid representations. For example, some chromosome may use binary coding and some use real-valued coding. Consequently, genetic algorithms are well suited for solve mixed-integer problems. Note that appropriate crossover and mutation operators should be applied for different representations. For example, 1-point crossover or uniform crossover may applied on binary-coded chromosome, and simulated binary crossover can be applied on real-coded chromosome.

1.2.3 Genetic Programming

Genetic algorithms can be used for solving combinatorial, integer, continuous or hybrid optimization problems. However, there are problems that do not fall in any of the above. For example, one wants to automatically generate an optimal decision tree for solving a classification problem; or one attempts to automatically generate a piece of programming code for accomplishing a particular tasks. In this case, it is no longer sufficient to use binary or real-coded coding to efficiently represent such problems.

Genetic programming was proposed by John Koza, who was also a student of John Holland in the late 1980s for evolving programming code, i.e., generating new code on the basis of some given code. To this end, a tree structure can be used to represent a piece of code. For example, as shown in Fig. 1.22, a piece of code of a loop can be represented by a tree structure. Similarly, given a number of primitive functions such as mathematical operators, one can use a tree structure to represent a function, as shown in Fig. 1.23.

The overall framework of genetic programming is similar to that of genetic algorithms. A genetic programming consists of the following main components. Starting

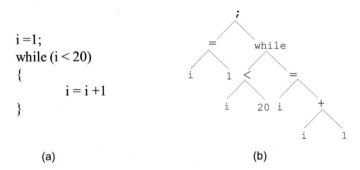

Fig. 1.22 An illustration of tree structure representation in genetic programming. The piece of code in (a) can be represented by the tree structure in (b)

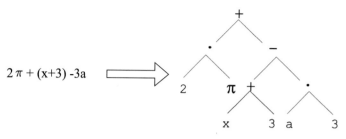

Fig. 1.23 Use of a tree structure to represent a function

from an initial parent population, crossover and mutation operations are applied to generate an offspring population. Then the next generation of parent population is generated by means of environmental selection such as proportionate selection or tournament selection. Due to the tree structure representation, crossover and mutation in genetic programming are adapted accordingly. To perform a crossover, a node for each parent will be selected and then the subtrees below the nodes will be exchanged. An illustrative example is provided in Fig. 1.24, where Parent 1 represents a function $sin(2x) + cos(y)$ and Parent 2 represents $cos(\pi + y)sin(x/5)$. After an exchange of the selected subtrees, the offspring individuals become $sin(2x) + cos(x/5)$ and $cos(y + \pi)sin(y)$.

To mutate a tree structure, a subtree of the individual will be randomly selected and replaced with a randomly generated subtree. It should be noted that in population initialization, crossover and mutation, one constraint should be paid attention to is the maximum depth of the trees to be generated, although the levels of depth may be different for different subtrees.

1.2.4   Evolutionary Multi-objective Optimization

The canonical evolutionary algorithms, including genetic algorithms and genetic programming, were developed for solving single-objective optimization problems as defined in Eq. (1.66). In the real world, however, most optimization problems have more than one objective, and these objectives are typically conflicting with each other, making it impossible to achieve the best value for each objective at the same time. For example, in portfolio optimization, one is interested to maximize the profit and minimize the risk. However, it is not likely to achieve an ideal solution that has the maximum profit and the minimum risk, as shown in Fig. 1.25. Instead, it is typical that the larger the profit, the higher the risk, i.e., there is a trade-off between profit and risk. Consequently, this bi-objective optimization problem does not have a single global optimum, like most single-objective optimization problem; instead, it has a set of trade-off solutions between profit and risk. Eventually, the user may select one of the solutions based on his or her preference. A cautious user may choose

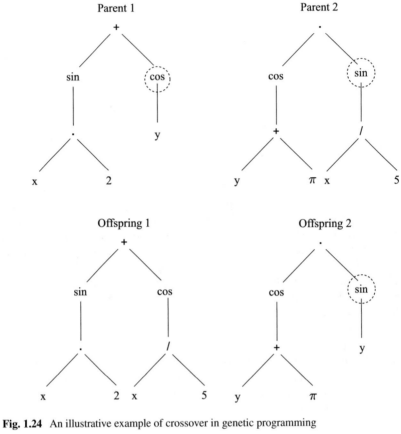

Fig. 1.24 An illustrative example of crossover in genetic programming

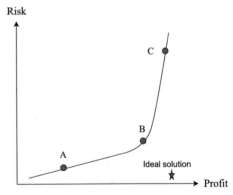

Fig. 1.25 A portfolio optimization problem, in which the profit (return) is to be maximized and the risk is to be minimized

solution A for a low risk and low return, while a more adventurous user may prefer solution C that has a high risk but also a high return.

Mathematically, a multi-objective optimization problem can be formulated as follows:

$$\text{minimize} \quad \{f_i(\mathbf{x})\} \qquad i = 1, 2, \ldots, m \tag{1.87}$$

$$\text{subject to} \quad g_j(\mathbf{x}) < 0, \quad j = 1, 2, \ldots, J. \tag{1.88}$$

$$h_k(\mathbf{x}) = 0, \quad k = 1, 2, \ldots, K. \tag{1.89}$$

$$\mathbf{x}^L \le \mathbf{x} \le \mathbf{x}^U, \tag{1.90}$$

where m is the number of objectives.

Before presenting ideas in mathematical programming and evolutionary computation for solving the above multi-objective optimization problem, we introduce an important concept known as Pareto dominance. A solution $\mathbf{x}_1$ is said to dominate another solution $\mathbf{x}_2$, if $f_j(\mathbf{x}_1) \le f_j(\mathbf{x}_2)$ for all $j = 1, 2, \ldots, m$, and there exists at least one $1 \le k \le m$ such that $f_k(\mathbf{x}_1) < f_k(\mathbf{x}_2)$. If a solution is not dominated by any other feasible solution, the solution is called Pareto optimal, and the image formed by all Pareto optimal solutions is called Pareto front or Pareto frontier.

Different multi-objective optimization problems have Pareto fronts of different nature. Fig. 1.26 shows four typical types of Pareto fronts, namely convex, concave, non-convex and discontinuous. The solution that has the maximum distance to the hyperplane (the line connecting two extreme solutions of the Pareto front) is called the knee point, e.g., solution A in Fig. 1.26(a), or solutions A and B in Fig. 1.26(c).

Usually, the dimension of the Pareto fronts equals $m - 1$. If the dimension of the Pareto fronts is smaller than $m - 1$, then this class of Pareto fronts are called degenerate. If the Pareto front does not span over the entire objective space, e.g., discontinuous or degenerate, it is called irregular.

A mathematical programming method typically converts a multi-objective optimization problem into a single-objective one before solving it. Different scalarization methods have been proposed and the simplest approach is the the linear weighted aggregation method that sums different objectives up with a set of weights to obtain a single objective:

$$F = \sum_{i=1}^{m} w_i f_i(\mathbf{x}), \tag{1.91}$$

where $0 \le w_i \le 1$, and $\sum_{i=1}^{m} = 1$.

The linear weighted aggregation approach has several weaknesses. First, a priori knowledge is needed to define the weights for the objectives beforehand. Second, one needs to run the optimization multiple times using different sets of weights if one is interested in obtaining multiple solutions. If the Pareto front is convex, different weight combinations will lead to different Pareto solutions, although evenly divided weights will not necessarily result in evenly distributed solutions. And even worse, for non-convex Pareto fronts, it is impossible to obtain the solutions in the concave regions, e.g., those between solutions A and B in Fig. 1.26(c), no matter how the weights are changed. Since the late 1990s, much research effort has been dedicated to the application of evolutionary algorithms to multi-objective optimization problems and great success has been accomplished. This might be attributed to the fact that evolutionary algorithms are population based, which are well suited for obtaining

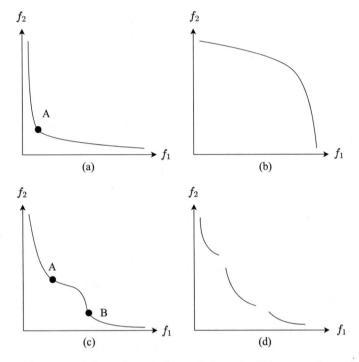

Fig. 1.26 Different types of Pareto fronts. **a** Convex, **b** Concave, **c** Non-convex, i.e., partly concave and partly convex, and **d** Discontinuous

a set of solutions in one single run. Roughly speaking, evolutionary algorithms for solving multi-objective optimization problems can be classified into three major categories.

The first category of evolutionary algorithms for multi-objective optimization is based on the Pareto dominance, based on which a rank is assigned to all solutions in the population and then perform environmental selection. Specifically, the rank of an individual equals the number of solutions that dominate it plus 1. That is, if a solution is not dominated by any other solutions in the current population, its rank is 1. Likewise, if a solution is dominated by two other solutions, then it will be assigned a rank of 3.

Given a rank assigned to each solution, a fitness value can then be calculated. Assume the rank of the i-th individual is $r_i, i = 1, 2, \ldots, N$, N is the population size, then the fitness of the i-th individual can be calculated as follows:

$$f_i = f_{\min} + (f_{\max} - f_{\min}) \frac{r_i - r_{\min}}{r_{\max} - r_{\min}}, \qquad (1.92)$$

where f_{max}, f_{min} are the user-defined maximum and minimum fitness values to be assigned, and r_{max} and r_{min} are the maximum and minimum ranks assigned to the individuals in the current population.

In principle, any environmental selection methods can be applied once each individual is assigned a fitness value. Recall, however, that the motivation of multi-objective optimization is to achieve a representative subset of the Pareto front. To achieve this, a diversity measure must be introduced to ensure the diversity of the solutions to be selected by biasing the assigned fitness values. For example, a niching count can be defined to penalize the fitness of an individual if there are solutions that are located very close to it. Let us define a threshold $d_{share} > 0$ and then the niche count of the i-th individual in the current population can be calculated as follows:

$$\mathrm{nc}_i = 1.0 - \sum_{j=1}^{N} \mathrm{Sh}_{ij}, \qquad (1.93)$$

where

$$\mathrm{Sh}_{ij} = \begin{cases} \frac{d_{ij}}{d_{share}}, & \text{if } d_{ij} \le d_{share} \\ 0, & \text{if } d_{ij} > d_{share} \end{cases} \qquad (1.94)$$

where d_{ij} is the Euclidean distance between the i-th individual and all others in the current population, and N is the population size.

From the above definition, we can see that the smaller the niche count, the more solutions there are near a solution, and the larger the penalty is. To illustrate the fitness assignment approach, we assume there are eight solutions in the population, as shown in Fig. 1.27. By comparing the dominance relationship between the seven solutions, we can see that solutions 1 and 2 are not dominated by any others, solutions 4, 6, and 7 are dominated by solution 2, solution 3 is dominated by solutions 1 and 2, solution 4 is dominated by solution 2, solutions 5 is dominated by solutions 1, 2, 3, 4, 6, and 7, and solution 8 is dominated by solutions 1, 2, 4, 6, 7. In addition, let $d_{share} = 0.5$, $d_{46} = 0.3$, $d_{64} = 0.3$, $d_{67} = 0.1$, $d_{76} = 0.1$, and all other $d_{ij} > 0.5$. In addition, let $f_{min} = 0$ and $f_{max} = 10$, then the rank, the assigned fitness, the niche count, and the shared fitness are listed in Table 1.1.

From the table, we can see that among solutions 3, 4, 6, and 7, the fitness value of solution 6 is most heavily penalized since it has two solutions very close to it, making it least likely selected among these four solutions to promote diversity.

The second category multi-objective optimization algorithms relies on the dominance relationship to rank solutions, based on which the environmental selection is performed. The most popular dominance based evolutionary algorithm for multi-objective optimization is known as the elitist non-dominated sorting genetic algorithm, NSGA-II for short [53]. A diagram of of NSGA-II is given in Fig. 1.28. From the diagram, we can see that NSGA-II is an elitism, since the best solutions in the parent population will be kept by combining the parent and offspring populations, unless all parent and offspring individuals are non-dominated. NSGA-II distinguishes with the single-objective genetic algorithm also in the following components. One is non-

Fig. 1.27 An illustrative
population consisting of
seven solutions

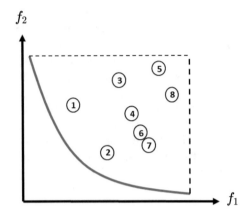

Table 1.1 The rank, niche count, assigned fitness, and shared fitness of the seven solutions

No.	Rank	Niche count	Fitness	Shared fitness
1	1	1	0	0
2	1	1	0	0
3	3	1	3.33	3.33
4	2	0.4	1.67	4.18
5	7	1	10	10
6	2	0.2	1.67	8.35
7	2	0.8	1.67	2.09
8	6	1	8.33	8.33

dominated sorting, which sorts the individuals in the combined population into a number of non-dominated fronts based on the dominance relationship. For example for the seven solutions in Fig. 1.29, solutions 1 and 2 are on the first non-dominated front, solutions 3, 4, 6, and 7 are on the second front, and solutions 5 and 8 are on the third non-dominated front. There are many different approaches to non-dominated sorting, including fast non-dominated sorting and efficient non-dominated sorting. These different sorting methods will result in the same result, and they distinguish mainly in computational complexity. A survey of different non-dominated sorting methods can be found in [54]. Once each individual is assigned a rank (front number), all individuals in the combined population are sorted in an ascending order according to the rank.

The second new component is the calculation of a crowding distance for each solution on the same non-dominated front, which is the average Euclidean distance in the objective values between two neighboring solutions on the same non-dominated front. For each of the two extreme solutions on each non-dominated front, a large crowding distance is assigned. In Fig. 1.29, for instance, the crowding distance of solution 4 is the measures by the distance between solutions 3 and 6. Accordingly,

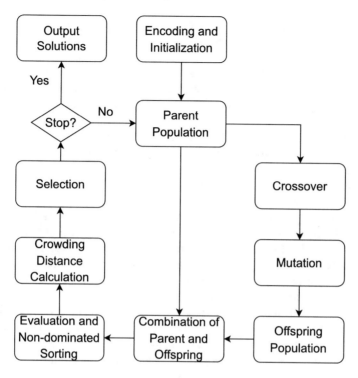

Fig. 1.28 A diagram of NSGA-II

Fig. 1.29 Non-dominated
sorting and crowding
distance calculation

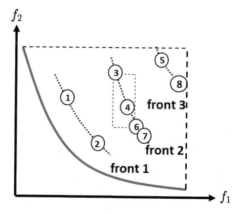

solutions 3 and 7 on the second front will be assigned a very large crowding distance so that they will be prioritized compared to solutions 4 and 6 in selection to promote diversity.

The crowding distance is used to indicate the degree of diversity of each solution, and the larger the crowding distance, the better the diversity performance the solution has. Once the crowding distance is calculated, the solutions on the same non-dominated front can be sorted in a descending order according to the crowding distance. For the eight solutions in Fig. 1.29, they are ranked in the following order: 1, 2, 3, 7, 4, 6, 5, 8. At this point, environmental selection can be carried out, which simply select the first half of the combined population as the parents of the next generation. For example, if we select four solutions from the above eight, 1, 2, 3, and 7 will be selected.

Evolutionary algorithms are population based search methods, and thus are well suited for multi-objective optimization as they can achieve a set of Pareto optimal solutions. One natural idea is therefore, to let each individual search for one Pareto optimal solution. Some early work proposed to use randomly generated weights [55] or dynamically changing weights [56]. In this case, an archive is usually needed to store the obtained non-dominated solutions. A more attractive approach is to define a set of evenly distributed weights so that evenly distributed Pareto optimal solutions can be obtained, assuming that the equally distributed weights can lead to a set of diverse solutions. A preliminary work along this line was reported in [57], where a cellular multi-objective genetic algorithm (C-MOGA) is proposed. In C-MOGA, N (the population size) equally distributed weights are defined:

$$F(\mathbf{x}) = \sum_{i=1}^{m} w_i f_i((x)), \tag{1.95}$$

$$\sum_{j=1}^{m} w_j = d. \tag{1.96}$$

where $i = 1, 2, \ldots, m$, m is the number of objectives. For example for a bi-objective optimization problem, if $d = 8$, then there will be 11 pairs of weights, namely, (0,8), (1,7), ..., (8, 0), as shown in Fig. 1.30(a). In this case, the population size is 9. For a three-objective optimization problem, let d=4, then there will be 16 pairs of weights, as shown in Fig. 1.30(b).

Given the predefined weights, one important issue is how to encourage the individuals, each of which optimizes on one sub single-objective problem, to work collaboratively. This is achieved by defining a neighborhood in the weight space, in which parents are randomly selected from one of the sub problem's neighborhood for generating its offspring by means of crossover and mutation. The fitness of the offspring will be re-calculated using the sub problem's weights. An improved version of C-MOGA, which is called multi-objective evolutionary algorithm based on decomposition (MOEA/D) [58]. MOEA/D has two main differences from C-MOGA. First, the parent individual of sub-problem is replaced by the offspring only if the

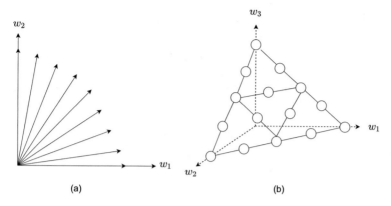

Fig. 1.30 Examples of equally distributed weights for **a** two-objective optimization, and **b** three-objective optimization. For the two-objective case, 9 pairs of weights are generated and for the three-objective case, 16 pairs of weights are generated

offspring is better than the parent. Second, a scalarizing function other than the linear weighted sum approach in (1.95) is suggested, which can effectively improve the performance on problems with a concave Pareto front. One popular scalarizing function is the Tchebycheff approach:

$$F_j(\mathbf{x}) = \max_{1 \le i \le m}\{\lambda_i^j | f_i(\mathbf{x}) - z_i^*|\} \tag{1.97}$$

where $\lambda^j = (\lambda_1^j, \lambda_2^j, \ldots, \lambda_N^j)$, N is the number of equally distributed weight vectors, typically also the population size, and z_i^* is reference point of the i-th objective, usually the best known value of this objective.

Evolutionary algorithms have achieved huge success in solving multi-objective optimization problems. While the early efforts have focused on two- and three-objective problems, increasing attention has been paid to problems with more than three objectives, which are known as many-objective optimization problems (MaOPs) [59]. For MaOPs, most evolutionary relying on dominance comparison will become less efficient, mainly because the number of non-dominated solutions within a limited population will quickly increase as the number of objectives increases, making the rank based selection less effective in driving the population to the Pareto front. To address this problem, either a modified dominance relationship should be adopted to increase the selection pressure, or an additional selection criterion needs to be introduced to distinguish non-dominated solution [60, 61]. Usually, decomposition based evolutionary algorithms still work for MaOPs, although a more elaborated scalarizing function will considerably enhance the performance, such as the reference vector guided evolutionary algorithm (RVEA) [62] that uses the angle as the distance measure rather than the Euclidean distance. Other issues such as handling irregularity in the Pareto front [63], and incorporating user preferences [64].

1.2.5   Evolutionary Multi-objective Learning

All machine learning problems are optimization problems, no matter whether they are supervised, unsupervised or reinforcement learning. This is easy to understand since each machine learning problem has a loss function to minimize. And as a matter of fact, a machine learning problem usually needs to consider more than one objective.

Take neural network regularization as an example. To mitigate the overfitting problem, an extra term describing the complexity of the neural network model will be added to the loss function:

$$L = E + \lambda\Omega, \tag{1.98}$$

where E is the error of the neural network on the training data, Ω is the model complexity, and $\lambda > 0$ is a hyperparameter. Typically, E is the mean squared error, and Ω is the sum of the squared weights

$$\Omega = \sum_{i=1}^{M} w_i^2, \tag{1.99}$$

or the sum of the absolute weights

$$\Omega = \sum_{i=1}^{M} |w_i|, \tag{1.100}$$

where $w_i, i = 1, 2, \ldots, M$ are all weights in the neural network. In traditional machine learning, the hyperparameter λ must be properly defined to train a neural network that can fit the training data while controlling the complexity of the model. Overfitting may still occur if λ is too small, whereas the model may underfit the data if λ is too large.

From the optimization point of view, the above neural network regularization problem us a typical bi-objective optimization problem, where minimizing the error on the training data and minimizing the model complexity are two conflicting objectives. Thus, it can be reformulated as follows [65]:

$$\min\{f_1, f_2\} \tag{1.101}$$
$$f_1 = E; \tag{1.102}$$
$$f_2 = \Omega. \tag{1.103}$$

In the above bi-objective minimization problem, no hyperparameter needs to be defined, and the solution set to the problem will be a set of Pareto optimal solutions, representing models having different complexities (Fig. 1.31).

Almost all machine learning problems have been approached from the Pareto optimization approach [65, 66]. A few examples are given in the following.

Fig. 1.31 Bi-objective
optimization of the neural
network regularization
problem, resulting in a set of
Pareto optimal models that
trade off between the
training accuracy and model
complexity

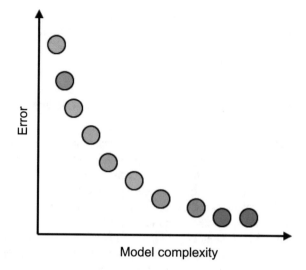

Model complexity

- Feature selection [67]. Feature selection can be seen as a bi-objective optimization problem, where the two objectives are the quality of the features and the number of the features.
- Feature extraction [68]. In extracting image features, minimization of the within class variance and maximization the between class variance of the features forms a bi-objective optimization problem.
- Data clustering [69]. Data clustering can also be formulated as a bi-objective optimization problem, where one objective is to maximize the inter-cluster similarity and the other is to minimize the intra-cluster similarity. One benefit of the Pareto approach to clustering is that one no longer needs to specify the number of clusters, which is a non-trivial task.
- Ensemble generation [70]. In ensemble generation, diversity of the base learners must be taken into account in addition to the accuracy of the ensemble.
- Interpretability enhancement of neural networks [65] and fuzzy rule systems [71]. For shallow neural networks, it has been shown that reducing the complexity of neural networks is helpful in enhancing their transparency in that understandable symbolic rules can be extracted from simplified neural networks [65]. Similarly, reducing the number of rules and the number of rule premises can lead to better interpretable fuzzy rules [71].
- Robustness to adversarial attacks [72]. To search for neural architectures that are robust to adversarial attacks on the data, a bi-objective optimization approach can be adopted that simultaneously maximizes the performance on the clean data and on the attacked data.
- Handling the trade-off between forgetting and memorizing [73]. Memorizing the old data and learning the new data, also known as catastrophic forgetting, can also be treated as a bi-objective optimization problem.

In Chap. 3 of this book, we will also present two multi-objective federated neural architecture search algorithms, aiming to maximizing the performance and minimizing the communication costs.

1.2.6 Evolutionary Neural Architecture Search

Neural architecture search (NAS) has become a hot research topic in recent years and it is able to automatically search the networks to outperform the ones manually designed by human experts. According to different search strategies, NAS is usually categorized into reinforcement learning (RL) based NAS, evolutionary based NAS, and gradient based NAS. In this section, evolutionary based approaches will be introduced and discussed due to its big success on NAS in recent years. Compared to other search strategies to perform NAS, population-based evolutionary algorithms (EAs) are less likely to be trapped into the local optimum and are naturally well suited to multi-objective optimization tasks.

A general process of evolutionary NAS is presented in Fig. 1.32 and it works like this: a population containing all the individual solutions are initialized at the beginning of the evolutionary optimization and each individual represents an encoded neural network architecture. And then, fitness values (e.g the model performance, the model size) of each architecture should be computed for selection (e.g tournament selection [74]) to generate the parent population in the evolutionary process. After that, crossover and mutation are applied to each individual of the parent population to generate the offspring population for fitness evaluations. Finally, environmental selection is adopted to generate the new parent population. This evolutionary process requires to be repeated until the predefined stop criteria is reached and the final population denotes the searched network architectures. Note that, the parent and offspring population can be combined together other than just the offspring for environmental selection to get solutions with possible better qualities and this kind of algorithm is often named as 'elitist' evolutionary algorithm [75].

Although the formal concept and the first RL-based NAS algorithm is proposed by Google research team in 2017 [76], evolving neural networks by EAs has already appeared in 1980s. And the original EA-based approaches use EAs to simultaneously optimize both model parameters and architectures [77–80]. However, this becomes less practical for modern deep neural networks containing millions or even billions of trainable parameters with complex structure. Consequently, recent evolutionary NAS algorithms only search or optimize the model architecture while leaving the model training to the fast and popular gradient descent methods [81].

Most evolutionary NAS algorithms [82] are designed for deep convolutional neural networks (CNNs) as they have shown superior abilities in the area of image classifications. As introduced in the previous section, a general CNN is consist of the convolutional layers, the pooling layers and the fully connected layers. Therefore, the architecture search space mainly includes the depth of neural layers, the hyperparameters of each layer and the neuron connections between each layer.

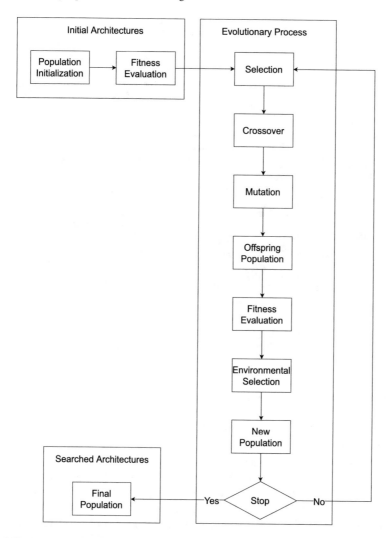

Fig. 1.32 A general process of evolutionary NAS

1.2.6.1 Search Space

The neural architectures are required to be encoded into individual chromosome in the population for EA-based NAS methods. And the encoding space directly determine the search space of NAS.

It is not wise to encode the whole CNN architecture layer by layer, leading to very large search space and long searching time. In order to avoid this issue, some constraints should be applied to the search space, resulting in marco search space and micro search space [83].

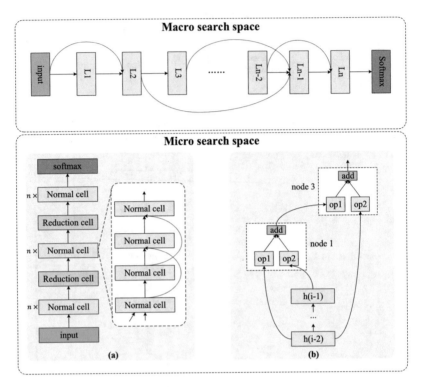

Fig. 1.33 Macro and Micro search space of NAS

As shown in Fig. 1.33, the macro search space aims to design the entire model structures consisting of n sequential blocks, model hyperparameters and the skip connections. Each rectangle block contains several different types of layers like the convolutional layers and some manually designed blocks like ResBlock [35], Inverted ResBlock [84] and so on. Since these blocks have already shown good performance in learning tasks, thus, requiring comparatively fewer parameters to build the entire architecture. However, the encoding flexibility still makes macro search space computationally expensive. And compared to marco search space, micro search space is a more efficient choice by searching the repeated normal cells and reduction cells (at bottom part of Fig. 1.33). The connections among these cells are usually fixed (this is not absolute), thus, significantly shrinking the search space and accelerating the search speed. In fact, micro search space has become the most popular search space in NAS.

Since the large amount of computational overhead is still at present a big barrier to the rapid development of NAS technologies, the search space is inadvisable to be set too large and the searched network architectures are sometimes relatively 'fixed'.

1.2.6.2 Encoding Strategy

Generally speaking, the encoding strategies can be categorized into two types: one is fixed-length encoding strategy and the other is variable-length encoding strategy [82]. Fixed-length encoding strategy makes all the chromosomes of individuals in the population have the same length, and it is intrinsically well suited to EA-based NAS approaches as the standard evolutionary operators like crossover and mutation require the lengths of two individuals to be the same. For instance, Genetic CNN [85] is one of the earliest studies using fixed-length encoding binary string to encode the entire architecture of the CNN model. Although this work has some limitations on the search scalability, the searched architecture can be transferred from CIFAR10 and SVHN datasets to ImageNet dataset.

By contrast, the individuals may have different chromosome lengths by adopting variable-length encoding strategy. Compared to the fixed-length approaches, variable-length encoding strategy [86] can encode more detailed information of the model architecture with freedom. For instance, the number of normal cells n in Fig. 1.33 are often fixed in the micro search space. However, if these numbers are undefined by a human expert, they should be encoded into the chromosome with a random integer number (can be binary numbers). It is possible that the optimal n can be found during the evolutionary optimization. However, variable-length encoding strategies are naturally not suited to the conventional evolutionary operators like crossover, and they may cause unexpected computational overhead if, for instance, the generated network architectures are very large.

1.2.6.3 One-Shot NAS

Although early research work with both fixed-length and variable-length encoding on evolutionary NAS are able to find high-quality architectures [87–89], these methods are overall very time consuming. And the new generated offspring population should be reinitialized and trained from scratch at each evolutionary generation. Note that the main computational bottleneck lies in evaluating the fitness of the individuals by invoking the lower-level weight optimization. One such evaluation typically takes several hours to days if the network is large or if the training dataset is huge. For instance, the AE-CNN [90] consumes 27 GPU days, CNN-GA [91] consumes 35 GPU days to search neural architectures on CIFAR10 dataset, while the large-scale evolutionary method AmoebaNet [86] consume terrible 3150 GPU days on the same data. This seriously limits the practical usability of most evolutionary NAS methods under a constrained search budget.

Furthermore, it is hard to deal with the trade-off relationship between the fitness precision and computational cost of each individual. For example, if each individual in the population adopts a large number of epochs to train the decoded model, a more accurate validation accuracy would be achieved for fitness evaluation, but must consuming large computational costs. Even though the population-based EAs are naturally suited to parallel computing, the computational overhead is still too heavy and unacceptable. By contrast, if a small number of training epochs are applied, it will be more likely to get an inaccurate validation accuracy, leading to unexpected evaluation biases for the evolutionary search.

Consequently, it is desired to redesign the topology of the search space to construct a more light weighted evolutionary optimization framework by using one-shot NAS methods. The core idea of one-shot NAS is to build a *supernet* search space containing all available architectures, and each individual is sub-sampled from the supernet. As a simple example shown in Fig. 1.34, where the supernet is actually a direct acyclic graph (DAG). Each node within the supernet represents a computational layer and the connected edges refer to the information flow between two layers. As a result, the supernet can directly inherit model parameters from each sampled subnet without training thousands of sub-models from scratch. The general steps of learning one-shot evolutionary NAS are summarized as follows:

1. Design a supernet model containing all possible architectures as the search space.
2. Train randomly sampled subnets to update the supernet until convergence by weight sharing technique [92].

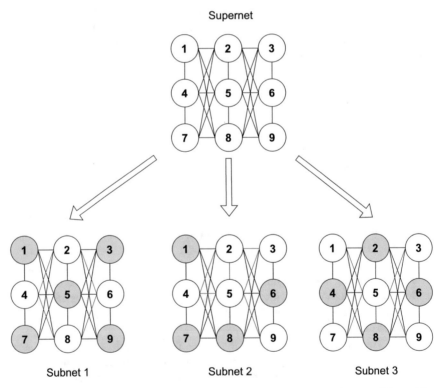

Fig. 1.34 An example of one-shot NAS, the shaded nodes denote inactive layers and the red line indicates activated path

3. Adopt a specific EA to find a brunch of subnets based on well trained supernet.
4. Train the subnet(s) (selected from the searched subnet set) from scratch to get the final trained network(s) with qualified architecture(s).

1.2.6.4 Multi-objective Optimization

As mentioned before, EA-based NAS methods are naturally well suited to deal with multi-objective problems. Instead of only considering one objective (e.g. the validation accuracy), other objectives like the number of parameters and floating-point operations per seconds (FLOPs) related to the computational efficiency are also simultaneously computed in fitness evaluations. Since these objectives are always in conflict with each other, multi-objective evolutionary algorithms like NSGA-II can be adopted for environmental selection [93] based on the dominate relationships of the individual solutions. Of course, multi-objective methods can be applied in the evolutionary process of one-short NAS.

1.2.6.5 Surrogate Assisted Evolutionary NAS

To further reduce the computational costs of the fitness evaluations and accelerate the search speed, surrogate assisted evolutionary algorithms (SAEAs) [94] are proposed to replace the computational expensive optimization problems (e.g. training DNNs for validation accuracy) into some cheap classification and regression models such as radial basis function networks (RBFNs) [95] and Gaussian process (GP) models [96, 97]. The general method is to do real evaluations on few individuals within a population, for example, performing expensive model training on a few number of candidate neural networks to get the real fitness values (e.g., validation accuracy). And then, the trained networks together with fitness values are used to build a surrogate model for cheap evaluations of remaining individuals. The basic steps of surrogate assisted evolutionary NAS are shown below:

1. Randomly sample a set of parent networks from the search space and train them (from scratch) to get the validation accuracy as data labels. The encoded architectures together with their corresponding labels are stored in an archive A.
2. Process archive A to construct the training data D_{tr} of the surrogate model M. Use D_{tr} to train and update M.
3. Use evolutionary operators (e.g. crossover and mutation) to generate the offspring population. Instead of training each individual model in the offspring, the validation performance are directly predicted by the surrogate model M.
4. Select candidate architectures based on the designed model management strategy to train them by the real data. The derived validation performance will be used to update A and D_{tr}. After that, the surrogate model can be upgraded by D_{tr}.

Note that the surrogate model is also named as a network performance predictor [98] for evaluating one objective value. And various surrogate models can be con-

Fig. 1.35 The workflow of
surrogate assisted
evolutionary NAS

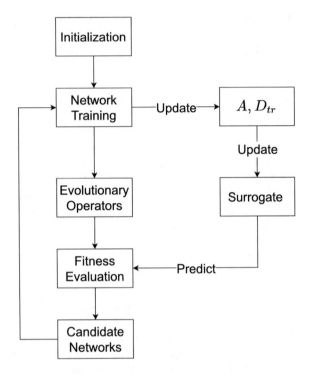

structed to deal with multi-objective optimization problems. The general workflow
of surrogate assisted evolutionary NAS is also shown in Fig. 1.35 and more details
related to this topic can be found in the book [99]. An overview of the methods for
reducing computational cost in NAS can be found in [100].

1.3 Privacy-Preserving Computation

Privacy preserving computation is mainly developed to protect sensitive data and provide secure information exchange among multiple parties. Technically, it includes
secure multi-party computation [101], differential privacy [102], homomorphic
encryption [103] and so on, which are widely used in the field of medical health,
finance and public transport. Of course, privacy preserving computation can be integrated with federated learning (FL) to further enhance the system security level. And
in this section, we would like to only introduce some practical privacy preservation
protocols suited to FL and other types of approaches are outside the scope of this
book.

1.3.1 Secure Multi-party Computation

The concept of secure multi-party computation (SMC) is first proposed in Yao's millionaire problem [104], and has been developed over several decades from pure theoretical research into the real-world applications [105]. Given N participants p_1, $p_2, \ldots p_N$, each containing the local data $D_1, D_2, \ldots D_N$. And the purpose of SMC is to collaboratively compute a public function $f(D_1, D_2, \ldots, D_N)$ without leaking any private information of the local data. One premise for the successful use of SMC is that all the participating parties should be *honest-but-curious*. That is all the parties strictly follow the process of the protocol but try to infer private information of others.

1.3.1.1 Garbled Circuit

The original Yao's protocol (also named garbled circuit) is the simplest cryptographic SMC algorithm to enable a secure two-party computation without the help of trusted third party (TTP). Assume that two individuals Alice and Bob want to privately compute the function $f(a, b)$, where $a \in \{0, 1\}$ and $b \in \{0, 1\}$ are private local data for Alice and Bob, respectively, and the computational circuit f (known to both Alice and Bob) can be represented by a simple example of the gate circuit as shown in Fig. 1.36.

If Alice is the garbler, she would garble (encrypt) the circuit gate by gate before sending it to the evaluator Bob. The garbling process performs like this: Alice builds the Truth table for every gate in the circuit, for instance, the logic gate located at the left top of Fig. 1.36 can be encoded into a Boolean circuit (Truth table) as shown in Fig. 1.37. And then, Alice replaces each Boolean element of the Truth table as a garbler-defined string (label). For example, the Boolean input a of Alice can be encoded into two randomly generated n-bit-length labels X_a^0 and X_a^1, respectively, and each which is also regarded as a key of symmetric encryption [106]. After that, Alice encrypts the encoded outputs X_c by two keys of the corresponding encoded

Fig. 1.36 An example of the gate circuit

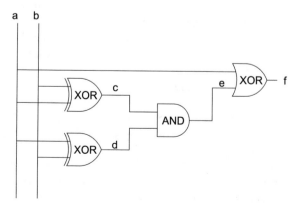

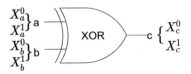

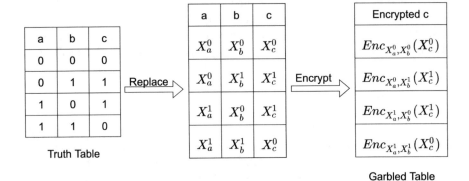

Fig. 1.37 An example of circuit garbling

gate inputs X_a and X_b as shown in the garbled table of Fig. 1.37. Each row of the garbled table requires to be permuted to hide the order of the table outputs before sending it to Bob, and the term 'garbled' is derived from this permutation.

After Bob receives the garbled tables of the entire circuit, he needs the input labels to decrypt the circuit. Therefore, Alice needs to send the labels of its data a to Bob. Assume that Alice's local data $a = 0$, she will send X_a^0 to Bob. Bob cannot get any information of a through received labels with format of a random string. In addition, Bob is not able to deduce a from the permuted garbled tables.

Bob also requires the labels of its own input data b as well, however, he has no knowledge of the encoding scheme for Alice. And it is strictly forbidden that Bob directly sends b to Alice due to the privacy concern. Consequently, oblivious transfer [107] can be adopted to fix this issue. Oblivious transfer is a type of protocols that, for example, can let Bob get the labels of its input b and Alice has no idea about b. If $b = 1$, Bob can get X_b^1 from $\{X_b^0, X_b^1\}$ on Alice, but Alice does not know which one is picked by Bob.

If Bob has already received the garbled tables and corresponding input labels, he can go through all gates one by one. Recall that, the garbled table is permuted, thus, Bob tries to decrypt the table row by row until the correct output is derived. For example, Bob has privately received X_a^0 and X_b^1, the encoded output c can be decrypted by equation $X_c^1 = Dec_{X_a^0, X_b^1}(Enc_{X_a^0, X_b^1}(X_c^1))$. Similar methods are adopted to sequentially decrypt the intermediate outputs of the gate circuit (Fig. 1.36): use X_a^0 and X_b^1 to decrypt X_d^1, use X_c^1 and X_d^1 to decrypt X_e^1, use X_a^0 and X_e^1 to decrypt and get the final encoded output X_f^1.

Finally, Bob sends X_f^1 to Alice and Alice can decode and share the output information to Bob. As a result, Both Alice and Bob get the computational output of the circuit without knowing each other's input data.

It is clear to see that SMC like garbled circuit can be easily applied in federated learning (FL) to further enhance the system security level. For a typical horizontal FL (HFL) problem, all N participants can use their local models $M_1, M_2, \ldots M_N$ to privately aggregate the global model $f(M_1, M_2, \ldots M_N)$. However, these kinds of SMC approaches are 'overqualified' for the global aggregation with just sum operations, and the complex cryptographic protocols must bring in extremely high computational costs to FL systems.

1.3.1.2 Practical Secure Aggregation

Recently, the Google researchers proposed a practical and efficient secure aggregation scheme [108] for HFL. Unlike most SMC algorithms [109–111] have very complex protocols with heavy encryption computations, practical secure aggregation method protects local privacy by adding random noise. And these added noise can be cancelled with each other after the global model aggregation on the server. A simplified process of secure aggregation protocol is shown as follows:

1. The central server generates a large prime number p, a generator g of cyclic group $\mathbb{G}$, a modular R and distributes these values to all connected N clients.
2. Each client i generates a random integer $sk_i \in \mathbb{Z}_p$ as a 'secret key' and computes $pk_i = g^{sk_i} \pmod{p}$. pk_i will be sent to any other client j, $j \neq i$, $j \in [N]$.
3. Each client i computes $pk_{j,i} = pk_j^{sk_i} \pmod{p}$ based on all received pk_j and generates the corresponding random noise $r_{j,i} = \alpha \cdot \mathrm{PRG}(t||pk_{j,i}) \pmod{R}$ where t is the communication round index, PRG is a selected pseudo random generator with seed $t||pk_{j,i}$ and α can be seen as an amplifier: if $i < j$, $\alpha = 1$, else $\alpha = 0$.
4. Each client i perturbs the updated local model (or gradients) M_i by the sum of all computed noise as $\tilde{M}_i = M_i + \sum_{j \neq i} r_{j,i}$, and sends $\tilde{M}_i$ to the server for aggregation.
5. The server simply sums all received $\tilde{M}_i$ to get the aggregated global model M (the perturbed noise will be canceled with each other).

It should be emphasized that any client j cannot get the secret key sk_i from the received pk_i due to discrete logarithm hard problem [112], even if g and p is known. Although this secure aggregation protocol requires to perform frequent Diffie' Hellman key exchange between each connected client, it is still the most efficient 'SMC' algorithm at present for HFL. Furthermore, the computational complexity of this protocol is robust to the number of model parameters. However, strictly speaking, secure aggregation protocol is not SMC method, since it only supports additive operations.

1.3.2 Differential Privacy

The concept of ϵ-differential privacy (DP) [102] is first proposed by the Microsoft research team in 2006. It is originally designed to protect the statistical privacy of the individuals in the publicly shared dataset. And the intuition behind DP is that any person's privacy will not be compromised no matter whether its data is in the database. In addition, the amount of any individual's contribution to the query result depends on the number of people in the query. For example, if the entire database contains data from n people, each person's contribution occupies $\frac{1}{n}$ of the query. This phenomenon indicates the key insight of differential privacy that more noise must be applied if the query is made on the data of fewer people. The definition of the standard ϵ-DP is shown below:

Definition (*ϵ-differential privacy* [113]). Given a real positive number ϵ and a randomized algorithm $\mathcal{A}$: $\mathcal{D}^n \rightarrow \mathcal{Y}$. Algorithm $\mathcal{A}$ provides ϵ—differential privacy, if for any two datasets $D, D^{'} \in \mathcal{D}^n$ differs on only one entity, and all $\mathcal{S} \subseteq \mathcal{Y}$ satisfy:

$$Pr[\mathcal{A}(D) \in \mathcal{S}] \leq exp(\epsilon) \cdot Pr[\mathcal{A}(D^{'}) \in \mathcal{S}] \tag{1.104}$$

where the Laplace mechanism is often adopted to satisfy the requirement of ϵ-DP by adding *Laplacian noise* to the deterministic function of the dataset $\mathcal{D}$. Based on the concept of ϵ-DP, a relaxation of ϵ, δ-DP named approximated DP [114] is defined as:

Definition (*(ϵ, δ)-differential privacy*). Given two real positive numbers (ϵ, δ) and a randomized algorithm $\mathcal{A}$: $\mathcal{D}^n \rightarrow \mathcal{Y}$. An algorithm $\mathcal{A}$ provides (ϵ, δ)-differential privacy if it satisfies:

$$Pr[\mathcal{A}(D) \in \mathcal{S}] \leq exp(\epsilon) \cdot Pr[\mathcal{A}(D^{'}) \in \mathcal{S}] + \delta \tag{1.105}$$

where δ denotes the probability for two adjacent datasets D and $D^{'}$ cannot be bounded by ϵ after adding a privacy preserving mechanism [115]. Different from ϵ-DP, Gaussian noise is instead applied to the output of the algorithm and the standard deviation σ of Gaussian noise satisfies $\sigma \geq \sqrt{2\ln(1.25/\delta)}$.

DP has an attribute of *composability*. For example, if we query an ϵ-DP for t times, and the randomization of the mechanism is independent of each query, and the result will be $t\epsilon$-deferentially private. Besides, the sensitivity of a function f denoted as Δf is given by Eq. (1.106):

$$\Delta f = \max \| f(D_1) - f(D_2) \|_2 \tag{1.106}$$

where D_1 and D_2 are all pairs of data in $\mathcal{D}$ differing in at most one element, and $\|\cdot\|_2$ represents ℓ_2 norm. And then, the Laplacian of Gaussian noise following $\mathcal{N}(0, \Delta f^2 \sigma^2)$ will be applied to the randomized function f.

According to the composability property, the noise of ϵ-DP achieves the complexity of $(O(q\epsilon), q\delta)$, where q is the sampling ratio per lot. To make the noise small while satisfying DP, the definition in [116] sets $\sigma \geq c \frac{q\sqrt{Tlog(1/\delta)}}{\epsilon}$, to achieve $(O(q\epsilon\sqrt{T}), \delta)$-DP where c is a constant, and T refers to the number of steps.

Next, we are going to introduce two types of ϵ, δ-DP usage in FL: one is called local differential privacy and the other is global differential privacy.

1.3.2.1 Global Differential Privacy

Global ϵ, δ-DP [117] is proposed to hide clients' contributions on the *server side* for defending membership inference [118] in FL systems. The core idea behind this method is to add noise upon the aggregated global model or gradients at each communication round t given by Eq. (1.107):

$$\theta_{t+1} = \theta_t + \frac{1}{m_t}\left(\sum_{k=0}^{m_t} \nabla\theta^k / \max\left(1, \frac{\|\nabla\theta^k\|_2}{\Delta f}\right) + \mathcal{N}(0, \Delta f^2\sigma^2)\right) \quad (1.107)$$

where θ_t is the global model parameters at the tth communication round, $\nabla\theta^k = \theta^k - \theta_t$ (can be approximated as the model gradients) of client k, m_t is the number of connected clients and $\mathcal{N}(0, \Delta f^2\sigma^2)$ is the added noise following the Gaussian distribution with 0 mean and $\Delta f^2\sigma^2$ variance.

Note that the Gaussian mechanism requires the knowledge about the set's sensitivity with respect to the summing operations. Since the updated model parameters θ^k are nondeterministic on each client k, it is hard to compute their boundaries, let alone the sensitivity Δf. Therefore, in order to satisfy ϵ, δ-DP, a manually defined sensitivity $\Delta f = \text{median}(\nabla\theta^k)_{k \in m_t}$ is determined by the median norm of all unclipped gradients $\nabla\theta^k$.

1.3.2.2 Local Differential Privacy

Local ϵ, δ-DP [119–121], on the other hand, requires each client k to perturb the updated local gradients $\nabla\theta^k$ with Gaussian noise before sending them to the central server where the local data information are hard to be deduced from the received uploads. The general step of federated local DP is shown below:

1. The server distributes the global model θ_t to each client k.
2. Each client k trains the downloaded model θ_t to get the updated local model θ^k and computes the difference $\nabla\theta^k = \theta^k - \theta_t$ as the model gradients which should be clipped into $g^k = \text{Clip}(\nabla\theta^k)$.
3. For local DP, each client k perturbs the clipped gradients g^k by the Gaussian noise into $\tilde{g}^k = g^k + \text{noise} \sim \mathcal{N}(0, \Delta f^2\sigma^2)$ and sends $\tilde{g}^k$ to the central server.
4. The server aggregates all the received $\tilde{g}^k$ to update the global model $\theta_{t+1} = \theta_t + \eta \sum_k \frac{n_k}{n} \tilde{g}^k$ where n_k is the training data size on the client k.
5. Repeat the above steps until the global model converges.

Overall, DP is a kind of noise-based privacy preservation technology commonly used in FL. Unlike aforementioned secure aggregation protocol, the perturbed noise generated by DP mechanism cannot be cancelled out with each other after the global model aggregation on the server. Therefore, DP exits a trade-off relationship between the global model performance and data privacy. In addition, the privacy budget is accumulated over the communication rounds and the training may be stopped at the early stage to cause underfitting if the privacy budge is beyond the threshold.

1.3.3   Homomorphic Encryption

Homomorphic encryption (HE) [103] is a type of asymmetric encryption algorithms to allow two or more cipertexts of the data encrypted with the same public key to do algebra computation directly. The computed ciphertext result, when decrypted, is equal to the value calculated with the same algebra operation on the plaintext data, making HE applicable to FL.

As a simple example of additive HE, two plaintexts m_1 and m_2 are encrypted by the same public key pk into two ciphertexts $\text{Enc}(m_1)$ and $\text{Enc}(m_2)$, respectively. It satisfies the equation $\text{Enc}(m_1) * \text{Enc}(m_2) = \text{Enc}(m_1 + m_2)$, where $*$ is the algebra operator like addition and multiplication. Similarly, multiplicative HE satisfies the condition of $\text{Enc}(m_1) * \text{Enc}(m_2) = \text{Enc}(m_1 \cdot m_2)$. In general, HE can be categorized into *partially* HE (PHE) and *Fully* HE (FHE).

1.3.3.1   Partially Homomorphic Encryption

PHE represents that the encrypted ciphertexts can only perform one kind of mathematical operation. Rivest' Shamir' Adleman (RSA) [122] is a classic multiplicative HE and is widely used in many real world applications like bank systems. It is an asymmetric crypotosystem containing two kinds of different keys: one is the public key pk and the other is the secret (private) key sk. The public key is used for encryption and can be shared to any others, while the secret key is used for decryption and should be kept privately.

The security of the RSA algorithm is based on the hardness of the factoring the product of two large prime numbers [123]. The RSA algorithm involves the following four steps:

1. **Key generation**: Choose two distinct large prime numbers p and q, and compute the modulus $n = pq$ whose bit length is regarded as the key length. And then, compute Carmichael's totient function $\lambda(n) = \text{lcm}(\lambda(p), \lambda(q))$, where $\lambda(p) = \varphi(p) = p - 1$. Hence, $\lambda(n) = \text{lcm}(p - 1, q - 1)$. After this, choose an integer e satisfying $1 < e < \lambda(n)$ and $\gcd(e, \lambda(n)) = 1$. Finally compute $d \equiv e^{-1}$ (mod $\lambda(n)$). Note that n and e are public keys, and d is the private key. In addition, p, q and $\lambda(n)$ should also be kept secret.

2. **Key distribution**: If party A wants to send information to party B, it will distribute the public keys n and e to party B for message encryption.
3. **Encryption**: Party B encodes its message M into an integer m ($0 \le m < n$) and encrypts m into the ciphertext c by equation $c = m^e \pmod{n}$. Party B sends c to party A.
4. **Decryption**: Party A decrypts the ciphertext c using the private key d by $c^d = (m^e)^d = m \pmod{n}$.

It is easy to prove that RSA is a multiplicative HE. Given two encoded messages m_1 and m_2, the corresponding encrypted ciphertexts will be $c_1 = m_1^e \pmod{n}$ and $c_2 = m_2^e \pmod{n}$, respectively. Apparently, $\text{Enc}(m_1) \cdot \text{Enc}(m_2) = m_1^e m_2^e = (m_1 m_2)^e \pmod{n} = \text{Enc}(m_1 \cdot m_2)$ satisfies the condition of multiplicative HE.

However, multiplicative HEs are not well suited to FL, since the global model aggregation only has addition operation. Therefore, additive HEs like Paillier encryption [124] are more commonly used in FL and Paillier works in the following way:

1. **Key generation**: Randomly select two large prime numbers p and q s.t. $\gcd(pq, (p-1)(q-1)) = 1$. Let $n = pq$ and $\lambda = \text{lcm}(p-1, q-1)$. After that, randomly choose an integer $g \in \mathbb{Z}_{n^2}^*$ and compute $\mu = (L(g^\lambda \bmod n^2))^{-1} \bmod n$, where L is a function defined as $L(x) = \frac{x-1}{n}$. The public key pk and secret key sk are (n, g) and (λ, μ), respectively.
2. **Encryption**: To encrypt a message $m \in \mathbb{Z}_n^*$, choose a random number $r \in \mathbb{Z}_n^*$ as an ephemeral key, the ciphertext is calculated as $c = g^m \cdot r^n \bmod n^2$.
3. **Decryption**: The plaintext message m can only be decrypted if the secret key (λ, μ) is available by computing $m = L(c^\lambda \bmod n^2) \cdot \mu \bmod n$.

The Paillier satisfies the additive homomorphic property: $\text{Enc}(m1) * \text{Enc}(m2) = \text{Enc}(m1 + m2) = g^{m_1} r_1^n \cdot g^{m_2} r_2^n \bmod n^2$.

Although additive HEs perfectly match FL from the perspective of algorithm level, the main challenges of combining HEs and FL often come from two aspects. Firstly, one secret key in FL systems may cause potential privacy leakage, since FL is a distributed machine learning scheme. For instance, an adversarial client may deliver the unique secret key to any other parties. This issue can be solved by distributed key sharing technologies like Shamir Secret Sharing [125] to divide the private key into several shards owned by n participating clients. The actual secret key can be recovered only if t out of n shards are collected.

Another challenge is the high computational costs caused by encryption computations. Without considering the engineering techniques like parallel computing, both encryption and decryption should be performed on each element of the model parameters one by one, causing unacceptable computational overhead. This problem becomes a disaster for deep learning models containing millions or even billions of parameters. And *batch encryption* [126] is one of possible approaches to accelerate HE speed by concatenating multiple plaintext vectors into a long vector with specific encoding methods. Thus, encrypting the encoded long vector is equivalent to encrypting multiple plaintexts in parallel, significantly improving the encryption efficiency.

Table 1.2 A comparison of Paillier, CKKS and batching versions of the previous two algorithms

# of plaintexts	HE	Ciphertext size	Encryption time (s)	Decryption time (s)	Addition time
	Paillier	8.00MB	20.02	11.52	5.41
	Paillier+batch	96.51 KB	0.46	0.37	0.06
	CKKS	6.60 GB	66.74	45.93	187.28
16384 (40.02 KB)	CKKS+batch	1.65 MB	0.02	0.01	0.05

1.3.3.2   Fully Homomorphic Encryption

FHE supports arbitrary computations on ciphertexts. The earliest formal FHE scheme is introduced in the research proposal of Gentry's PhD thesis based on ideal lattices [127]. After that, many researchers pay their attention to this exciting research topic. Dijk et al. [128] propose a simple FHE with elementary modular arithmetic based on Gentry's work. This method only adopts addition and multiplication on integers other than using ideal lattices, which is similar to the framework proposed in [129].

More recently in 2017, Cheon, Kim, Kim and Song (CKKS) [130] propose a FHE system for approximate arithmetic, supporting approximate addition and multiplication on cipertexts. In addition, CKKS adopts a novel rescaling method to manage the magnitude of the encrypted data. Note that most FHE cryptosystems like BGV [131] only work on integers, thus, parameters of ML models (always real numbers) require to be encoded into integers before doing encryptions on them.

CKKS has been applied to FL in recent research work [132]. Unlike aforementioned additive HE cryptosystems such as Paillier encrypting model parameters element by element, CKKS is able to directly encrypt the plaintext polynomial as a whole. And the polynomial is encoded by the plaintext vector converted from the model parameters. In addition, batch encryption can be combined with CKKS to further reduce the encryption (decryption) time, addition time and ciphertext size [133]. A comparison of Paillier, CKKS and their batch encrypted versions is shown in Table 1.2. It is clear to see that batch encryption can significantly reduce the computational costs for both Paillier and CKKS methods.

1.4   Federated Learning

1.4.1   Horizontal and Vertical Federated Learning

Generally speaking, federated learning can be divided into two main categories: horizontal federated learning and vertical federated learning [134]. Detailed discussions are provided below.

1.4.1.1   Horizontal Federated Learning

Horizontal federated learning, also known as homogeneous federated learning [135], represents scenarios where the clients' datasets have the same feature space but different sample spaces. As shown in Fig. 1.38, the row data represents the personal information of a user, including age, gender, height and weight, and corresponding labels. It can be seen that the local datasets of client 1 and client 2 share the same feature space, but the user groups are different.

In a standard horizontal federated learning system, the global model and all local models have the same model structure, and only differ in model parameters. In general, setting local epochs E, local batch size B and learning rate η appropriately according to the task may improve the performance of the federated learning system to a certain extent. Compared with centralized learning, horizontal federated learning provides a simple and effective solution to prevent private local data leakage, because only global model parameters and local model parameters are allowed to communicate between the server and clients, and all training samples are saved on the client device. As a result, they are inaccessible to any other client.

However, frequently downloading and uploading model parameters consumes a lot of communication resources, and the situation becomes more serious if the task itself requires a lot of computing and memory resources. Therefore, researchers have proposed various strategies to reduce the communication costs, such as client update subsampling and model quantization.

1.4.1.2   Vertical Federated Learning

Vertical federated learning [134, 136] is also known as heterogeneous federated learning [137], which represents the scenarios where the clients' feature spaces are different but the sample space is the same. As shown in Fig. 1.39, client 1 and client 2 have the same data samples but different features. In contrast to horizontal federated learning, where each client has their own local data labels, in vertical federated

Features

Name	Age	Sex	Height	Weight	Label
Person A	24	Male	178	78	1
Person B	61	Female	165	64	0
Person C	44	Male	182	89	1
Person D	17	Female	159	52	0
Person E	11	Male	137	36	1
Person F	33	Female	171	60	0

Client 1 (Person A–C), Client 2 (Person D–F), Samples

Fig. 1.38 An example of horizontal federated learning

Features

Name	Age	Height	Label
Person A	24	178	1
Person B	61	165	0
Person C	44	182	1
Person D	17	159	0
Person E	11	137	1
Person F	33	171	0

Name	Sex	Weight
Person A	Male	78
Person B	Female	64
Person C	Male	89
Person D	Female	52
Person E	Male	36
Person F	Female	60

Samples

Client 1   **Client 2**

Fig. 1.39   An example of vertical federated learning

learning, only one client is believed to retain all data labels (e.g. in Fig. 1.39, only client 1 has data labels). The client with labels is called *guest* client [138] or *active* client [139], while clients without data labels are called *host* clients or *passive* clients.

In general, the labels of the samples must be the same for all clients, otherwise predictions cannot be made correctly. Besides training data, there are three main differences between horizontal and vertical federated learning. First, horizontal federated learning has a server for global model aggregation, while vertical federated learning has neither a server nor a global model. Therefore, the main process of vertical federation is to perform model prediction on the *host* client, upload it to the *guest* client for aggregation and then construct the loss function. Second, model parameters or gradients are passed between the server and clients in horizontal federated learning. But in vertical federated learning, local model parameters are not allowed to be transferred to other clients. Instead, the *guest* client receives the model output from the connected *host* client and sends back intermediate gradient values for local model updates. Finally, the server and clients only interact once in one round of communication in horizontal federated learning, while in a communication round between *guest* client and *host* client in vertical federated learning, information needs to be sent and received multiple times.

Compared to horizontal federated learning, vertical federated learning has two advantages on model training. First, a model trained in vertical way should in principle have the same performance as a model trained centrally, because the loss function computed in vertical federated learning is the same as that in centralized learning [140]. Second, vertical federated learning usually consumes less communication resources than horizontal federated learning, because only local model output and intermediate gradients need to be transmitted between *guest* client and *host* clients, therefore, the communication cost in vertical federated learning depends on the number of data samples, so only when the amount of data is very large, vertical federated learning may consume more communication resources than horizontal federated learning.

1.4.2 Federated Averaging

The increasing privacy concerns and data protection intentions lead to the emergence of privacy-protected distributed learning methods, and federated learning, as one of the most representative algorithms, enables multiple clients collaboratively learn a global model with satisfied performance under heterogeneous data distributions has received much attention in recent years. Federated averaging (FedAvg) [141], as one of the most prevalent methods for model aggregation, plays a key role in building federated learning systems. As the name implies, FedAvg is the final step of each communication round that aims to average all received local models from the participated clients, which can be expressed by

$$w_t = \sum_{k=1}^{\lambda K} p_k w_t^k, \tag{1.108}$$

where t is the index of communication round, λ indicates the participation ratio and K is the total number of clients consisted in the system, and $p_k = \frac{n_k}{n}$ ($\sum p_k = 1$) reflects the ratio of the number of local samples n_k to the number of all samples n. Intuitively, FedAvg emphasises on the aggregation of local models, and an equivalent implementation of FedAvg is FedSGD, in which each client conducts *one local iteration* per communication round,

$$w_t = w_{t-1} - \eta g_{t-1}^{(1)} \tag{1.109}$$

$$= w_{t-1} - \eta \sum_{k=1}^{\lambda K} p_k g_{t-1}^{k(1)} \qquad /*FedSGD*/ \tag{1.110}$$

$$= \sum_{k=1}^{\lambda K} p_k w_{t-1} - \eta \sum_{k=1}^{\lambda K} p_k g_{t-1}^{k(1)} \tag{1.111}$$

$$= \sum_{k=1}^{\lambda K} p_k (w_{t-1} - \eta g_{t-1}^{k(1)}) \tag{1.112}$$

$$= \sum_{k=1}^{\lambda K} p_k w_t^k \qquad /*FedAvg*/, \tag{1.113}$$

where $g_{t-1}^{k(1)}$ is the gradient of local client k at round t for 1 local iteration (epoch).

Compared to FedSGD, FedAvg proposes two main motivations: First, as we mentioned above, the authors think Eq. (1.110) is equivalent to Eq. (1.108). Second, they allow additional local iterations per round to reduce communication costs in federated learning systems, which is

$$w_{t-1}^{k(i)} = w_{t-1}^{k(i)} - \eta g_{t-1}^{k(i)}, \quad 0 \le i \le E, \quad w_{t-1}^{k(0)} = w_{t-1}, \tag{1.114}$$

where E is a hyper-parameter usually set to 5.

Nevertheless, *Eq. (1.110) does not hold when the number of local iterations exceeds 1*, according to [142], since *the assumption* $\sum p_k g_{t-1}^{k(i)} = g_{t-1}^{(i)} (i \geq 1)$ *holds if and only if* $g_{t-1}^{k(i)}$ *is extracted from the same data distribution*. To prove this, we can have a glance at the gradients in the second local iteration:

$$g_{t-1}^{(2)} = w_{t-1}^{(1)} - \eta g_{t-1}^{(1)}, \tag{1.115}$$

$$g_{t-1}^{k(2)} = w_{t-1}^{k(1)} - \eta g_{t-1}^{k(1)}, \tag{1.116}$$

if the data distribution on client k varies from the global data distribution, $w_{t-1}^{k(1)}, g_{t-1}^{k(1)}$ will be different from $w_{t-1}^{(1)}, g_{t-1}^{(1)}$, and the error will accumulate as the optimization progresses. To address this issue, a variety of methods have been proposed. For example, Yao et al. introduced an unbiased gradient aggregation method to keep trace the global gradient descent direction [142].

Interestingly, FedAvg averages all local models element-wise according to there coordinates or indices, instead of averaging elements with "the same function". This strategy does work when the model is large or deep enough, but fails to converge for traditional machine learning models. For example, a mismatch problem may occur when average two radial basis function network (RBFN) [143], which can be easily observed through the following toy experiment in Fig. 1.40.

To avoid the mismatch problem, Xu et al. [143] average the parameters according to their roles in the networks and propose a sorted averaging algorithm. The intuition is simple but effective, which can be illustrated by Fig. 1.41.

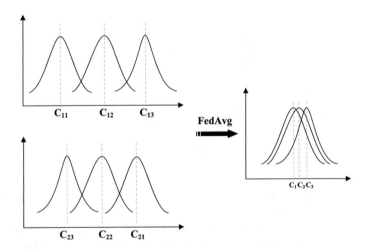

Fig. 1.40 An illustrative example of aggregating two local univariate RBFNs, each having three nodes. The three Gaussian functions of the local RBFNs are shown on the left, and the resulting three Gaussian functions of the aggregated global RBFN are plotted on the right

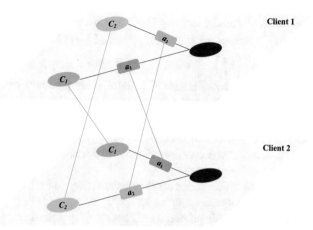

Fig. 1.41 Sorted averaging. C_i, a_i indicate the i_{th} center and corresponding weight of a local RBFN, respectively. The authors use Euclidean distance to sort the index of centers and average them accordingly

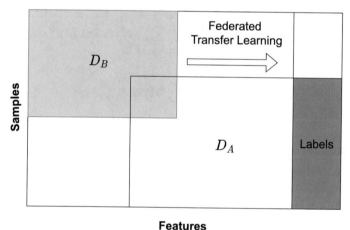

Fig. 1.42 Federated transfer learning

1.4.3 Federated Transfer Learning

Federated transfer learning is first proposed in [136] to describe the scenarios in which the client data differ not only in feature space but also in sample space [134]. For two financial agents, one is the bank located in the UK and the other is securities company located in China. Due to geographical and business differences, these two agents have dramatically different user groups and provide different services. A simple example of a two-party FTL is described in Fig. 1.42, where D_A and D_B are the training data on client A and client B, respectively.

For the definition of transfer learning [144], given a binary source domain dataset $D_A := \{(x_i^A, y_i^A)\}_{i=1}^{N_A}$ where $x_i^A \in R^a$ and $y_i^A \in \{+1, -1\}$ is the i-th label. A target domain dataset $D_B := \{x_j^B\}_{j=1}^{N_B}$ where $x_j^B \in R^a$. For a simple two-party federated transfer learning system, two datasets D_A and D_B are privately located on client A and client B. It is also assumed that these two clients have a limited set of co-occurrence data samples $D_{AB} := \{x_i^A, x_i^B\}_{i=1}^{N_{AB}}$ and a small set of labels for target domain data D_B in client A's dataset $D_c := \{x_i^B, y_i^A\}_{i=1}^{N_c}$. If not specified, all the training data labels are stored on client A (source domain), which is similar to the case of VFL.

Then, two hidden representations for the ith data entry on both client A and client B can be computed by equations $u_i^A = M^A(x_i^A)$ and $u_i^B = M^B(x_i^B)$, respectively, where M^A and M^B are both private local models (e.g. neural networks) used to extract input features into common subspace.

The bridge of knowledge transfer comes from two paths: one is the co-occurrence data samples D_{AB} and the other is D_c with a small set of available labels. In order to label the target domain, the prediction $\varphi(u_j^B) = \varphi(u_1^A, y_1^A, ...u_{N_A}^A, y_{N_A}^A, u_j^B)$ is computed to construct the loss. If $\varphi(u_j^B)$ is linearly separable, it can be divided into two parts $\varphi(u_j^B) = \Phi^A G(u_j^B)$, where $\Phi^A = \frac{1}{N_A} \sum_{i=1}^{N_A} y_i^A u_i^A$ and $G(u_j^B) = (u_j^B)'$. Thus, the training loss can be constructed into Eq. (1.117):

$$\underset{\theta^A, \theta^B}{\arg\min} \; \mathcal{L}_1 = \sum_{i=1}^{N_c} \ell_1(y_i^A, \varphi(u_i^B)) \tag{1.117}$$

where θ^A and θ^B are model parameters of local models M^A and M^B, respectively, and ℓ_1 is, for example, the logistic loss function $\ell_1(y, \varphi) = \log(1 + e^{-y\varphi})$. Moreover, the feature transfer learning is achieved by minimizing the distance between u^A and u^B on the aligned datasets D_{AB} as shown in Eq. (1.118), where ℓ_2 denotes the alignment loss such as $-u_i^A(u_i^B)'$ and $\|u_i^A - u_i^B\|_2$.

$$\underset{\theta^A, \theta^B}{\arg\min} \; \mathcal{L}_2 = \sum_{i=1}^{N_{AB}} \ell_2(u_i^A, u_i^B) \tag{1.118}$$

The total loss can be computed by combing $\mathcal{L}_1$ and $\mathcal{L}_2$ into Eq. (1.119), where γ and λ are weight hyperparameters, and $\|.\|_2$ is the regularization term.

$$\underset{\theta^A, \theta^B}{\arg\min} \; \mathcal{L} = \mathcal{L}_1 + \gamma \mathcal{L}_2 + \frac{\lambda}{2}(\|\theta^A\|_2 + \|\theta^B\|_2) \tag{1.119}$$

After the total loss is constructed, the gradients for updating local model θ^i, $i \in \{A, B\}$ can be easily derived by Eq. (1.120):

$$\frac{\partial \mathcal{L}}{\partial \theta^i} = \frac{\partial \mathcal{L}_1}{\partial \theta^i} + \gamma \frac{\partial \mathcal{L}_2}{\partial \theta^i} + \lambda \theta^i \tag{1.120}$$

It is easy to find that constructing the total loss in Eq. (1.119) requires the private information from both client A and client B. Therefore, the gradients of the total loss with respect to local models should be securely calculated due to the privacy preservation requirements of federated transfer learning. One popular approach is to adopt additive homomorphic encryption (HE) (denoted as $[[.]]$) upon the loss function, and the gradients are computed with the format of ciphertexts. At the same time, a second order Taylor approximation is used to simplify the original loss function, making the encrypted computations much easier. The approximated gradients with additive HE for both θ^A and θ^B are given by Eqs. (1.121) and (1.122), respectively.

$$
\left[\left[\frac{\partial \mathcal{L}}{\partial \theta^A}\right]\right] = \sum_{j=1}^{N_A} \sum_{i=1}^{N_c} (\frac{1}{4} D(y_i^A) \Phi^A \left[\left[\mathcal{G}(u_i^B)' \mathcal{G}(u_i^B)\right]\right]
$$
$$
+ \frac{1}{2} C(y_i^A) \left[\left[\mathcal{G}(u_i^B)\right]\right]) \frac{\partial \Phi^A}{\partial u_j^A} \frac{\partial u_j^A}{\partial \theta^A} \tag{1.121}
$$
$$
+ \gamma \sum_{i=1}^{N_{AB}} \left([[ku_i^B]] \frac{\partial u_i^A}{\partial \theta^A} + \left[\left[\frac{\partial \ell_2^A(u_i^A)}{\partial \theta^A}\right]\right] \right) + [[\lambda \theta^A]]
$$

$$
\left[\left[\frac{\partial \mathcal{L}}{\partial \theta^B}\right]\right] = \sum_{i=1}^{N_c} \frac{\partial \left(\mathcal{G}(u_i^B)\right)' \mathcal{G}(u_i^B)}{\partial u_i^B} \left[\left[\left(\frac{1}{8} D(y_i^A)(\Phi^A)' \Phi^A\right)\right]\right] \frac{\partial u_i^B}{\partial \theta^B}
$$
$$
+ \sum_{i=1}^{N_c} \left[\left[\frac{1}{2} C(y_i^A) \Phi^A\right]\right] \frac{\partial \mathcal{G}(u_i^B)}{\partial u_i^B} \frac{\partial u_i^B}{\partial \theta^B} \tag{1.122}
$$
$$
+ \sum_{i=1}^{N_{AB}} \left([[\gamma k u_i^A]] \frac{\partial u_i^B}{\partial \theta^B} + \left[\left[\gamma \frac{\partial \ell_2^B(u_i^B)}{\partial \theta^B}\right]\right] \right) + [[\lambda \theta^B]]
$$

where $C(y) = -y$, $D(y) = y^2$, and k is the constant. These two equations look really complex, however, they can be simplified into two parts: one is the terms can be computed locally on each client and the other is the encrypted terms from another client. As a simple example for $[[\frac{\partial \mathcal{L}}{\partial \theta^A}]]$, the term $\frac{1}{4} D(y_i^A) \Phi^A$ can be easily calculated by the local training data (x_i^A, y_i^A) on client A, while the encrypted term $[[\mathcal{G}(u_i^B)' \mathcal{G}(u_i^B)]]$ is computed on client B. Although there exists exceptions like $\left[\left[\frac{\partial \ell_2^A(u_i^A)}{\partial \theta^A}\right]\right]$ in $[[\frac{\partial \mathcal{L}}{\partial \theta^A}]]$, this method gives a really good intuitive understanding for these two equations.

Consequently, if we define the encrypted components on client A as $\left\{h_k(u_i^A, y_i^A)\right\}_{k=1,\ldots K_A}$, and in this simple case, $K_A = 3$, $h_1(u_i^A, y_i^A) = \left\{\left[\left[\frac{1}{8}D(y_i^A)(\Phi^A)'\Phi^A\right]\right]\right\}_{i=1}^{N_c}$, $h_2(u_i^A, y_i^A) = \left\{\left[\left[\frac{1}{2}C(y_i^A)\Phi^A\right]\right]\right\}_{i=1}^{N_c}$, and $h_3(u_i^A, y_i^A) = \left\{\left[\left[\gamma k u_i^A\right]\right]\right\}_{i=1}^{N_{AB}}$. Similarly, the encrypted components on client B can be defined as $\left\{h_k(u_i^B, y_i^B)\right\}_{k=1,\ldots K_B}$. And the training steps of HE-base FTL are shown below:

1. Client A and client B create their local models M^A and M^B, and randomly generate key pairs of additive HE. The generated public keys are exchanged with each other.
2. Client A computes u_i^A and sends encrypted $\left\{h_k(u_i^A, y_i^A)\right\}_{k=1,\ldots K_A}$ to client B. Similarly, client B A computes u_i^B and sends encrypted $\left\{h_k(u_i^B)\right\}_{k=1,\ldots K_B}$ to client A.
3. Client A creates a random mask r^A and computes $\left[\left[\frac{\partial \mathcal{L}}{\partial \theta^A} + r^A\right]\right]_B$ (subscript B means it is encrypted by the public key from client B) using the received ciphertexts $\left\{h_k(u_i^B)\right\}_{k=1,\ldots K_B}$. Similarly, client B generates a random mask r^B and computes $\left[\left[\frac{\partial \mathcal{L}}{\partial \theta^B} + r^B\right]\right]_A$ using received $\left\{h_k(u_i^A, y_i^A)\right\}_{k=1,\ldots K_A}$.
4. Client A sends $\left[\left[\frac{\partial \mathcal{L}}{\partial \theta^A} + r^A\right]\right]_B$ to client B, client B decrypts and sends $\frac{\partial \mathcal{L}}{\partial \theta^A} + r^A$ back to client A. Client B sends $\left[\left[\frac{\partial \mathcal{L}}{\partial \theta^B} + r^B\right]\right]_A$ to client A, client A decrypts and returns $\frac{\partial \mathcal{L}}{\partial \theta^B} + r^B$ back to client B.
5. Client A updates the local model $\theta^A = \theta^A - \eta \frac{\partial \mathcal{L}}{\partial \theta^A}$. Client B updates the local model $\theta^B = \theta^B - \eta \frac{\partial \mathcal{L}}{\partial \theta^B}$.
6. Go back to the step 2 and repeat this process until convergence.

It should be emphasized that $\frac{\partial \mathcal{L}}{\partial \theta}$ requires to be perturbed by locally generated random mask r, so that the gradients will not be leaked to other clients after decryption. For more FTL protocols with secret sharing, the reader is referred to [136].

Overall, above introduced FTL approach is close to VFL, since all the data labels are assumed to be located at only one client and intermediate representation outputs $[[u]]$ other than the model weights are transferred between two clients to construct the loss function on the 'guest' client. In addition, FTL makes an assumption that distributed clients exist aligned data samples D_{AB} and D_c for feature and label knowledge transfer, respectively, implying the term 'transfer learning' in FL.

1.4.4 Federated Learning Over Non-IID Data

Although FL can effectively protect users' local private data during distributed model training, models trained in a federated environment often underperform those trained in the standard centralized learning mode, especially when the training data are not independent and identically distributed (Non-IID) on the local devices. More specifically, for a supervised learning task on client k, assume each data sample (x, y), where x is the input attribute or feature and y is the label, follows a local

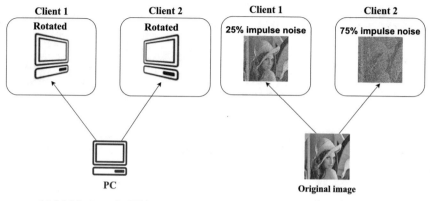

(a) Multi-view of a PC image.

(b) Images perturbed by different levels of impulse noise.

Fig. 1.43 Two examples of attribute distribution skew

distribution $P_k(x, y)$. By Non-IID, we mean P_k differs from client to client, which can usually be categorized into attribute skew and label skew [140].

1.4.4.1 Attribute Skew

Attributes, also known as features or characteristics, usually serve as the input or decision domain x of a machine learning model. Correspondingly, attribute skew indicates the scenarios in which the feature distribution $P_k(x)$ across attributes on each client is different from each other, and it varies from non-overlapped to fully overlapped skew corresponding the scenarios from VFL to HFL.

Non-overlapping attribute skew means that the data features across the clients are mutually exclusive. In this case, if data samples x on different clients k with the same identities hold the same labels $P_k(y|x)$, it is known as VFL. A simple example is shown in Fig. 1.39 that client 1 holds the data features of age and height, while client 2 owns the features of weight and sex.

Partially overlapping attribute skew denotes parts of the data features across different clients are overlapped and shared with each other. Multi-view images [145, 146] shot from different angles on different clients can be regarded as a type of partial overlapping skew. As shown in Fig. 1.43(a), client 1 and client 2 own the same PC image with different angles.

Fully overlapping attribute skew often refers to the data distribution situation in HFL, partitioning the overall dataset 'horizontally' along the sample direction (shown in Fig. 1.38). One kind of this attribute imbalance for the training data across connected clients is caused by different levels of perturbations. As a simple example shown in Fig. 1.43(b), two images on client 1 and client 2 are perturbed with different degrees of impulse noise. On the other hand, feature imbalance is another feature

Fig. 1.44 Label size
imbalance

	bird	deer	frog	ship
Client 1				
Client 2				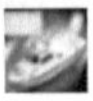

distribution skew with the same data features. For and written digit numbers from different people, thus, even for the same digit number, the character features (e.g., stroke width and slant) are different.

1.4.4.2   Label Skew

Label skew represents the scenarios in which the label distribution differs from client to client. In general, there are two different kinds of label skew, one is label distribution skew and the other is label preference skew.

Label distribution skew denotes the label distributions $P_k(y)$ vary among the clients while the conditional feature distribution $P_k(x|y)$ is shared across the clients. And this is usually caused by geographic variations of client devices storing similar types of datasets. Label distribution skew can be further categorized into label size imbalance and label distribution imbalance.

Label size imbalance refers to the data partition methods introduced in the original FL paper [147], in which each client owns data samples with a fixed number c label classes. c is a hyperparameter that determines the degree of label imbalance, and a smaller c means stronger label imbalance, and vice versa. An illustrative example of label size imbalance with the number of label classes $c = 2$ using CIFAR10 dataset is shown in Fig. 1.44.

Label distribution imbalance [148] can be seen as a generalized version of label size imbalance. The number of instances for label class c is instead controlled by a Dirichlet distribution $p_c \sim Dir_k(\beta)$ [149–152], where β is a concentration parameter affecting the imbalance level. And a smaller β value will result in more unbalanced data partition.

Label preference skew considers the client data sample *intersection* issues where the conditional distribution $P_k(y|x)$ may vary across the clients, even $P_k(x)$ is the same. For instance, it is likely that different users annotate different labels for the same data sample due to individuals' preferences. Furthermore, crowdsourcing data [153] is a more complex scenarios of label preference skew. It is assumed that most client devices contain a large amount of training data labels by multiple workers or volunteers.

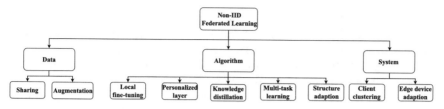

Fig. 1.45 A summary of existing approaches to addressing Non-IID data

1.4.4.3   Main Approaches to Non-IID Data

The original FedAvg algorithm [147] is not robust to model divergence caused by Non-IID data in FL, especially when the training models are complex containing a large amount of parameters (e.g. deep neural networks). Current methods dealing with Non-IID issues can be categorized into the data based approach, algorithm based approach, and system based approach, as shown in Fig. 1.45. Note that these Non-IID methods are mainly adopted upon training *parametric* models in HFL, since other cases such as VFL are not easy to be affected by Non-IID data [140]. Consequently, if not specified otherwise, FL means HFL, the baseline FL algorithm is FedAvg, and learning models are neural networks.

1.4.4.4   Data Based Approaches

Data based approaches are the most straight-forward methods to alleviate the side effect of Non-IID data. The core idea is to 'fill up' the Non-IID client data and make them become 'even' or 'IID' by data sharing or data augmentation techniques.

Data sharing [154–156] is the simplest but most powerful method that can outperform most FL algorithms on Non-IID data. It assumes that the central server owns a set of IID global data which can be publicly shared to distributed clients. And then, these shared (or partially shared) global data can be trained together with the local data to significantly enhance the model performance. However, data sharing significantly violates the privacy preservation requirement of FL that data cannot be shared and exchanged between the server and clients.

Data augmentation [157], on the other hand, increases the local data diversity to mitigate the data imbalance issues. Three types of data augmentation methods are usually adopted in FL: the vanilla method [158], the mixup method [159], and the generative adversarial network (GAN) [160] based method.

The vanilla augmentation method requires each client to upload its label distribution information to the server in which the overall data distribution can be calculated by combining the received information. After that, the server can tell each client the number of data should be generated (augmented) for class c. Both the original local data and the augmented data are used for training.

The mixup method assumes that there exists base data samples on the server and each client uploads its seed data samples [161] encoded by the XOR operator. And then the decoded samples are combined with the base samples to construct a new balanced dataset. Thus, the global model can be trained with reconstructed 'IID' dataset to alleviate the negative influence on local Non-IID data.

The GAN method augments the local data by creating a private generator on each client. And the shared global model is now becomes the discriminator of the GAN. By alternatively updating the trainable parameters of the shared discriminator and local generators, the generator on each client is able to generate synthetic data belonging to locally unavailable classes.

Overall, data based approaches can significantly improve the learning performance of the model trained on Non-IID data, however, they may cause data privacy leakage, more or less, during the training period.

1.4.4.5 Algorithm Based Approaches

Algorithm based approaches denote those methods implementing various machine learning algorithms to solve Non-IID problems in FL. As shown in Fig. 1.45, algorithm based approaches consist of local fine tuning, personalization layers, and knowledge distillation.

Local fine tuning refers to the methods redesigning the model updating rules on the client side. It aims to fully utilize the information difference between the local and global model, and tries to shrink the local training biases caused by Non-IID data. One common idea for fine-tuning is to interpolate the regularization term [162] upon the loss function used in the standard FedAvg algorithm as shown in Eq. (1.123):

$$f_k(\theta^k) = \frac{1}{n_k}\sum_i^{n_k} \ell(\boldsymbol{x}_i, y_i; \theta^k) + \frac{\gamma}{2}||\theta^k - \theta_t||^2 \qquad (1.123)$$

where θ_t is the global model at the t-th communication round and θ^k is the local model on client k. By optimizing this fine-tuned local loss, the disparity between the global and local models will be reduced.

Personalization layers means each client owns some 'personal' neural network layers kept privately on the local device without being shared with others. As a typical paradigm named FedPer [163] shown in Fig. 1.46, where the base layers (blocks depicted by solid lines) are the shallow layers of the neural networks that extracts high-level representations, and the personalization layers (blocks depicted by dashed lines) are the deep layers for classifications. The base layers are shared between the server and connected clients for collaborative learning, while the personalization layers are stored locally to preserve local trait. As a result, the base layers together with the local personalization layers not only learn generalized knowledge from the global model but also have good focus to deal with local Non-IID data.

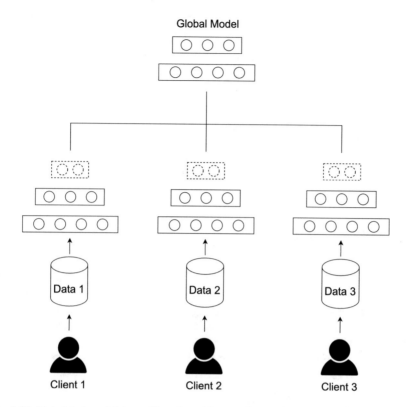

Fig. 1.46 Pictorial view of federated learning with personlization layers

Multi-task learning methods convert FL into multi-task problems where each task represents one local training process. For example, the goal of a representative multi-task framework MOCHA [164] is to consider solving distributed primal-dual optimization [165–167] challenges such as communication costs, stragglers and fault tolerance for FL.

Knowledge distillation has recently become a popular technique to solve Non-IID issues in FL. It is originated from the work in [168] that transfers the knowledge information from the large 'teacher' model to the small 'student' model. When adopting knowledge distillation in FL, the server is assumed to contain some unlabelled data and its general training process (e.g., FedDF [169]) is summarized in Algorithm 2, where AVGLOGITS is the global model updating rule given by Eq. (1.124):

$$\theta_{t,j} = \theta_{t,j-1} - \eta \frac{\partial \mathrm{KL}\left(\sigma(\frac{1}{|S_t|}\sum_{k\in S_t} f(\theta_t^k, \boldsymbol{d})), \sigma(f(\theta_{t,j-1}, \boldsymbol{d}))\right)}{\partial \theta_{t,j-1}} \quad (1.124)$$

Algorithm 2 An illustration of the homogeneous FedDF.

1: Initialize server model θ_0, K clients, a total of T communication rounds, n_k samples for a local
   dataset of client $k \in K$ and its weight p_k, participation ratio C.
2: **for** each communication round $t = 1, 2, \ldots T$ **do**
3:     $S_t \leftarrow$ random subset (C fraction) of K clients
4:     **for** each client $k \in S_t$ **in parallel do do**
5:         $\theta_t^k \leftarrow$ Local update of FedAvg using θ_{t-1}^k
6:     **end for**
7:     initialize for model fusion $\theta_{t,0} \leftarrow \sum_{k \in S_t} p_k \theta_t^k$
8:     **for** j in $\{1, \ldots, N\}$ **do**
9:         sample a mini-batch of samples d, from e.g., (1) an unlabeled dataset, (2) a generator using
         ensemble of $\{\theta_t^k\}_{k \in S_t}$ to update the server student $\theta_{t,j-1}$ through AVGLOGITS
10:    **end for**
11:    $\theta_t \leftarrow \theta_{t,N}$
12: **end for**
13: **Return** θ_T

where KL is the Kullback' Leibler divergence, σ is the softmax activation function
and η is the learning rate. Apparently, the teacher model is the ensembled client model
$\{\theta_t^k\}_{k \in S_t}$ and the student model is the aggregated global model $\theta_{t,j-1}$, and federated
knowledge distillation is performed by minimizing the KL divergence between the
outputs of the ensemble model and the global model for N iterations.

Structure adaptation refers to the methods adjusting training optimizers or network layers to accelerate the convergence speed of deep neural networks (DNNs) in
FL. For instance, adopting adaptive optimizers such as Adagrad [170], Adam [171]
and etc. to replace the standard SGD optimizer for faster learning speed on Non-IID
data. Moreover, the commonly used batch normalization layers [172] in deep neural
networks can be replaced by group normalization layers [33] to alleviate the learning
biases of batch moving statistics caused by Non-IID data.

1.4.4.6 System Based Approaches

System based approaches deal with Non-IID problems from the perspective of system
management and optimization. They are mainly comprised of client clustering and
edge device adaption.

Client clustering is proposed to construct a multi-center framework by grouping
the clients into different clusters. And in this case, the whole FL systems contain
various global models, each of which is aggregated by the client models within the
same cluster. The motivation is that one global model does not have enough 'capacity'
to learn all the client information, especially in a heterogeneous data environment.

The core idea of client clustering is to make secure data similarity measurement,
and those clients with similar local training data are expected to be assigned into
the same cluster for model aggregation. Specifically, data similarity can be evaluated through, for example, the local loss values computed upon the received cluster
(global) models from the server [173–176]. And then each client can determine

its corresponding cluster with the smallest loss value for subsequent cluster model aggregation. Since those 'unsimilar' client models within different groups are not aggregated, each cluster model will not be 'polluted' by models trained with heterogeneous datasets.

Edge device adaption is to design the incentive-based interaction between a global server and participating clients to determine the participation of devices in FL [177]. For example, the server can act like a leader to dynamically set the training hyperparameters such as the number of communication rounds, the number of local epochs to maximize the FL performance on Non-IID data according to the clients' responses.

In summary, FL over Non-IID data remains a big challenge. Most existing approaches such as local fine-tuning and data sharing achieve better convergence performance at the cost of either increased computational costs or even data privacy leakage. Other methods like client clustering need to change the structure of the vanilla FL framework, making it no longer possible to generate a global model for all clients. These problems should be solved in future research work.

1.5 Summary

This chapter provides the fundamentals of the book, including the basics of artificial neural networks, evolutionary optimization and learning, single- and multi-objective optimization and learning, privacy-preserving computing and federated learning. The main challenges of federated learning, i.e., learning over non-iid data are discussed in greater detail. In the next chapter, we will present two algorithms for reducing the communication costs in federated learning, one based on layer-wise asynchronous learning and temporally weighted aggregation, and the other using ternary compression.

References

1. Menabrea, L.F., Lovelace, A.: Sketch of the analytical engine invented by charles babbage. Sci. Mem. **3** (1843)
2. McCulloch, W.S., Pitts, W.: A logical calculus of the ideas immanent in nervous activity. Bull Math Biophys **5**, 115–133 (1943)
3. Hebb, D.O.: The Organization of Behavior. Wiley, New York (1949)
4. Rosenblatt, F.: The perceptron: a probabilistic model for information storage and organization in the brain. Psychol Rev **65**(6), 386–408 (1958)
5. Turing, A.M.: I.-Computing machinery and intelligence. Mind **236**, 433–460 (1950)
6. Weizenbaum, J.: ELIZA-A computer program for the study of natural language communication between man and machine. Commun ACM **9**, 136–45 (1966)
7. Nof, S.Y.: Handbook of Industrial Robotics. Wiley, New York (1999)
8. Zadeh, L.A.: Fuzzy sets. Inf. Control **8**(3), 338–353 (1965)
9. Schwefel, H.P.: Kybernetische evolution als strategie der exprimentellen forschung in der strömungstechnik. Master's thesis, Technical University Berlin (1965)

10. Minsky, M., Papert, S.: Perceptrons: An Introduction to Computational Geometry. The MIT Press, Cambridge MA (1969)
11. Werbos, P.J.: Beyond regression: New tools for prediction and analysis in the behavioral sciences. Ph.D. thesis, Harvard University, Cambridge MA (1974)
12. Holland, J.: Adaptation in Natural and Artificial Systems. University of Michigan Press, Ann Arbor (1975)
13. Hans-Paul, S.: Evolution strategy and numerical optimization (in german). Ph.D. thesis, Technical University of Berlin, Berlin (1974)
14. Kohonen, T.: Self-organized formation of topologically correct feature maps. Biol. Cybern. **43**, 59–69 (1982)
15. Barto, A.G., Sutton, R.S., Anderson, C.W.: Neuronlike elements that can solve difficult learning control problems. IEEE Trans. Syst. Man Cybern. **13**, 835–846 (1983)
16. Rumelhart, D., Hinton, G., McClelland, J.: A general framework for parallel distributed processing. In: Rumelhart, D., McClelland, J., The PDP Research Group (eds.) Parallel Distributed Processing: Explorations in the Microstructure of Cognition, vol. 1. The MIT Press, Cambridge, MA (1986)
17. Bienenstock, E., Cooper, L.N., Munro, P.: Theory for the development of neuron selectivity: orientation specificity and binocular interaction in visual cortex. J Neurosci **2**, 32–48 (1982)
18. Bi, G.Q., Poo, M.M.: Synaptic modifications in cultured hippocampal neurons: dependence on spike timing, synaptic strength, and postsynaptic cell type. J. Neurosci. **18**(24), 10464–10472 (1998)
19. Maass, W.: Networks of spiking neurons: the third generation of neural network models. Neural Netw. **10**(9), 1659–1671 (1997)
20. Fukushima, K., Miyake, S., Ito, T.: Neocognitron: a neural network model for a mechanism of visual pattern recognition. IEEE Trans. Syst. Man Cybern. **13**(3), 826–834 (1983)
21. LeCun, Y., Bottou, L., Bengio, Y., Haffner, P.: Gradient-based learning applied to document recognition. Proc. IEEE **86**(11), 2278–2324 (1998)
22. Cortes, C., Vapnik, V.: Support-vector networks. Mach. Learn. **20**, 273–297 (1995)
23. Hochreiter, S., Schmidhuber, J.: Long short-term memory. Neural Comput. **9**(8), 1735–1780 (1997)
24. Engelbrecht, A.P.: Computational Intelligence: An Introduction. Wiley, New York (2007)
25. Hinton, G.E.: Learning multiple layers of representation. TRENDS Cogn. Sci. **11**(10), 428–434 (2007)
26. Pham, D.: Neural networks in engineering. WIT Trans. Inf. Commun. Technol. **6** (1970)
27. Pal, S.K., Mitra, S.: Multilayer Perceptron, Fuzzy Sets, Classifiaction (1992)
28. Gardner, M.W., Dorling, S.: Artificial neural networks (the multilayer perceptron)-a review of applications in the atmospheric sciences. Atmos. Environ. **32**(14–15), 2627–2636 (1998)
29. Kanal, L.N.: Perceptron, pp. 1383–1385. Wiley, GBR (2003)
30. Rodríguez, O.H., Lopez Fernandez, J.M.: A semiotic reflection on the didactics of the chain rule. Math. Enthus. **7**(2), 321–332 (2010)
31. Singh, D., Singh, B.: Investigating the impact of data normalization on classification performance. Appl. Soft Comput. **97**, 105524 (2020)
32. Ioffe, S., Szegedy, C.: Batch normalization: accelerating deep network training by reducing internal covariate shift. In: International Conference on Machine Learning, pp. 448–456. PMLR (2015)
33. Wu, Y., He, K.: Group normalization. In: Proceedings of the European Conference on Computer Vision (ECCV), pp. 3–19 (2018)
34. Glorot, X., Bordes, A., Bengio, Y.: Deep sparse rectifier neural networks. In: Gordon, G., Dunson, D., Dudík, M. (eds.) Proceedings of the Fourteenth International Conference on Artificial Intelligence and Statistics, Proceedings of Machine Learning Research, vol. 15, pp. 315–323. PMLR, Fort Lauderdale, FL, USA (2011). https://proceedings.mlr.press/v15/glorot11a.html
35. He, K., Zhang, X., Ren, S., Sun, J.: Deep residual learning for image recognition. In: Proceedings of the IEEE Conference on Computer Vision and Pattern Recognition, pp. 770–778 (2016)

36. Yeh, I.C., Lien, C.h.: The comparisons of data mining techniques for the predictive accuracy of probability of default of credit card clients. Expert Syst. Appl. **36**(2), 2473–2480 (2009)
37. Moro, S., Cortez, P., Rita, P.: A data-driven approach to predict the success of bank telemarketing. Decis. Support. Syst. **62**, 22–31 (2014)
38. De Boer, P.T., Kroese, D.P., Mannor, S., Rubinstein, R.Y.: A tutorial on the cross-entropy method. Ann. Oper. Res. **134**(1), 19–67 (2005)
39. Harris, D., Harris, S.L.: Digital Design and Computer Architecture. Morgan Kaufmann (2010)
40. Bottou, L.: Stochastic gradient descent tricks. In: Neural Networks: Tricks of the Trade, pp. 421–436. Springer, Berlin (2012)
41. Everitt, B.S., Skrondal, A.: The Cambridge Dictionary of Statistics (2010)
42. Warde-Farley, D., Goodfellow, I.J., Courville, A., Bengio, Y.: An empirical analysis of dropout in piecewise linear networks. arXiv:1312.6197 (2013)
43. Srivastava, N., Hinton, G., Krizhevsky, A., Sutskever, I., Salakhutdinov, R.: Dropout: a simple way to prevent neural networks from overfitting. J. Mach. Learn. Res. **15**(56), 1929–1958 (2014). http://jmlr.org/papers/v15/srivastava14a.html
44. Crow, F.C.: Summed-area tables for texture mapping. In: Proceedings of the 11th Annual Conference on Computer Graphics and Interactive Techniques, pp. 207–212 (1984)
45. Hochreiter, S., Schmidhuber, J.: Long short-term memory. Neural Comput. **9**(8), 1735–1780 (1997). https://doi.org/10.1162/neco.1997.9.8.1735
46. Olver, P.J.: Applications of Lie Groups to Differential Equations, vol. 107. Springer Science & Business Media (2000)
47. Chen, T., Guestrin, C.: Xgboost: A scalable tree boosting system. In: Proceedings of the 22nd ACM SIGKDD International Conference on Knowledge Discovery and Data Mining, KDD '16, pp. 785–794. Association for Computing Machinery, New York, NY, USA (2016). https://doi.org/10.1145/2939672.2939785
48. Quinlan, J.R.: Induction of decision trees. Mach. Learn. **1**(1), 81–106 (1986)
49. Goldberg, D.E.: Genetic Algorithms in Search, Optimization, and Machine Learning. Addison Wesley (1989)
50. Herrera, F., Lozano, M., Sanchez, A.M.: A taxonomy for the crossover operator for real-coded genetic algorithms: an experimental study. Int. J. Intell. Syst. **18**, 309–339 (2003)
51. Deb, K., Agrawal, R.B.: Simulated binary crossover for continuous search space. Complex Syst. **9**(2), 115–148 (1995)
52. Hansen, N., Ostermeier, A.: Completely derandomized self-adaptation in evolution strategies. Evol. Comput. **9**(2), 159–195 (2001)
53. Deb, K., Pratap, A., Agarwal, S., Meyarivan, T.: A fast and elitist multiobjective genetic algorithm: NSGA-II. IEEE Trans. Evol. Comput. **6**(2), 182–197 (2002)
54. Tian, Y., Wang, H., Zhang, X., Jin, Y.: Effectiveness and efficiency of non-dominated sorting for evolutionary multi- and many-objective optimization. Complex Intell. Syst. **3**(4), 247–263 (2017)
55. Jin, Y., Okabe, T., Sendhoff, B.: Adapting weighted aggregation for multiobjective evolution strategies. In: Proceedings of the First International Conference on Evolutionary Multi-Criterion Optimization, pp. 96–110 (2001)
56. Jin, Y., Olhofer, M., Sendhoff, B.: Dynamic weighted aggregation for evolutionary multi-objective optimization: why does it work and how? In: Proceedings of Genetic and Evolutionary Computation Conference, pp. 1042–1049 (2001)
57. Murata, T., Ishibuchi, H., Gen, M.: Specification of genetic search directions in cellular multi-objective genetic algorithms. In: Proceedings of the First International Conference on Evolutionary Multi-Criterion Optimization, pp. 82–95 (2001)
58. Zhang, Q., Li, H.: MOEA/D: a multiobjective evolutionary algorithm based on decomposition. IEEE Trans. Evol. Comput. **11**(6), 712–731 (2007)
59. Li, B., Li, J., Tang, K., Yao, X.: Many-objective evolutionary algorithms: a survey. ACM Comput. Surv. **48**, 13–35 (2015)
60. Zhang, X., Tian, Y., Jin, Y.: A knee point driven evolutionary algorithm for many-objective optimization. IEEE Trans. Evol. Comput. **19**(6), 761–776 (2015)

61. Deb, K.: Multi-objective optimization. In: Search Methodologies, pp. 403–449. Springer, Berlin (2014)
62. Cheng, R., Jin, Y., Olhofer, M., Sendhoff, B.: A reference vector guided evolutionary algorithm for many-objective optimization. IEEE Trans. Evol. Comput. **20**(5), 773–791 (2016)
63. Hua, Y., Liu, Q., Hao, K., Jin, Y.: A survey of evolutionary algorithms for multi-objective optimization problems with irregular pareto fronts. IEEE/CAA J. Automatica Sinica **8**(2), 303–318 (2021)
64. Yu, G., Ma, L., Jin, Y., Du, W., Liu, Q., Zhang, H.: A survey on knee-oriented multi-objective evolutionary optimization. IEEE Trans. Evol. Comput. (2022)
65. Jin, Y., Sendhoff, B.: Pareto-based multiobjective machine learning: an overview and case studies. IEEE Trans. Syst. Man Cybern. Part C (Applications and Reviews) **38**(3), 397–415 (2008). https://doi.org/10.1109/TSMCC.2008.919172
66. Jin, Y. (ed.): Multi-objective Machine Learning. Springer, Berlin (2006)
67. Bing Xue, W.F., Zhang, M.: Multi-objective feature selection in classification: a differential evolution approach. In: Asia-Pacific Conference on Simulated Evolution and Learning, pp. 516–528 (2014)
68. Albukhanajer, W.A., Briffa, J.A., Jin, Y.: Evolutionary multi-objective image feature extraction in the presence of noise. IEEE Trans. Cybern. **45**(9), 1757–1768 (2015)
69. Handl, J., Knowles, J.: Exploiting the trade-off—the benefits of multiple objectives in data clustering. In: Third International Conference on Evolutionary Multi-Criterion Optimization, pp. 547–560. Springer, Berlin (2005)
70. Gu, S., Cheng, R., Jin, Y.: Multi-objective ensemble generation. WIREs Data Min. Knowl. Discov. **5**(5), 234–245 (2015)
71. Wang, H., Kwong, S., Jin, Y., Wei, W., Man, K.: A multi-objective hierarchical genetic algorithm for interpretable rule-based knowledge extraction. Fuzzy Sets Syst. **149**, 149–186 (2005)
72. Liu, J., Jin, Y.: Multi-objective search of robust neural architectures against multiple types of adversarial attacks. Neurocomputing **453**, 73–84 (2021)
73. Jin, Y., Sendhoff, B.: Alleviating catastrophic forgetting via multi-objective learning. In: International Joint Conference on Neural Networks, pp. 6367–6374. IEEE (2006)
74. Miller, B.L., Goldberg, D.E., et al.: Genetic algorithms, tournament selection, and the effects of noise. Complex Syst. **9**(3), 193–212 (1995)
75. King, R., Rughooputh, H.: Elitist multiobjective evolutionary algorithm for environmental/economic dispatch. In: The 2003 Congress on Evolutionary Computation, 2003. CEC '03., vol. 2, pp. 1108–1114 (2003). https://doi.org/10.1109/CEC.2003.1299792
76. Zoph, B., Le, Q.V.: Neural architecture search with reinforcement learning. In: 5th International Conference on Learning Representations, ICLR 2017, Toulon, France, Conference Track Proceedings. OpenReview.net (2017). https://openreview.net/forum?id=r1Ue8Hcxg
77. Schaffer, J., Whitley, D., Eshelman, L.: Combinations of genetic algorithms and neural networks: a survey of the state of the art. In: [Proceedings] COGANN-92: International Workshop on Combinations of Genetic Algorithms and Neural Networks, pp. 1–37 (1992). https://doi.org/10.1109/COGANN.1992.273950
78. Yao, X.: Evolving artificial neural networks. Proc. IEEE **87**(9), 1423–1447 (1999). https://doi.org/10.1109/5.784219
79. Stanley, K.O., Miikkulainen, R.: Evolving neural networks through augmenting topologies. Evol. Comput. **10**(2), 99–127 (2002). https://doi.org/10.1162/106365602320169811
80. Inden, B., Jin, Y., Haschke, R., Ritter, H.: Evolving neural fields for problems with large input and output spaces. Neural Netw. **28**, 24–39 (2012)
81. Bengio, Y.: Practical recommendations for gradient-based training of deep architectures. In: Neural Networks: Tricks of the Trade, pp. 437–478. Springer, Berlin (2012)
82. Liu, Y., Sun, Y., Xue, B., Zhang, M., Yen, G.G., Tan, K.C.: A survey on evolutionary neural architecture search. In: IEEE Transactions on Neural Networks and Learning Systems, pp. 1–21 (2021). https://doi.org/10.1109/TNNLS.2021.3100554
83. Pham, H., Guan, M., Zoph, B., Le, Q., Dean, J.: Efficient neural architecture search via parameters sharing. In: Dy, J., Krause, A. (eds.) Proceedings of the 35th International Conference

on Machine Learning, Proceedings of Machine Learning Research, vol. 80, pp. 4095–4104. PMLR (2018). https://proceedings.mlr.press/v80/pham18a.html

84. Sandler, M., Howard, A., Zhu, M., Zhmoginov, A., Chen, L.C.: Mobilenetv2: Inverted residuals and linear bottlenecks. In: Proceedings of the IEEE Conference on Computer Vision and Pattern Recognition, pp. 4510–4520 (2018)

85. Xie, L., Yuille, A.: Genetic cnn. In: Proceedings of the IEEE International Conference on Computer Vision, pp. 1379–1388 (2017)

86. Real, E., Moore, S., Selle, A., Saxena, S., Suematsu, Y.L., Tan, J., Le, Q.V., Kurakin, A.: Large-scale evolution of image classifiers. In: Precup, D., Teh, Y.W. (eds.) Proceedings of the 34th International Conference on Machine Learning, Proceedings of Machine Learning Research, vol. 70, pp. 2902–2911. PMLR (2017). https://proceedings.mlr.press/v70/real17a.html

87. Miikkulainen, R., Liang, J., Meyerson, E., Rawal, A., Fink, D., Francon, O., Raju, B., Shahrzad, H., Navruzyan, A., Duffy, N., Hodjat, B.: Chapter 15—evolving deep neural networks. In: Kozma, R., Alippi, C., Choe, Y., Morabito, F.C. (eds.) Artificial Intelligence in the Age of Neural Networks and Brain Computing, pp. 293–312. Academic (2019). https://doi.org/10.1016/B978-0-12-815480-9.00015-3, https://www.sciencedirect.com/science/article/pii/B9780128154809000153

88. Liang, J., Meyerson, E., Miikkulainen, R.: Evolutionary architecture search for deep multitask networks. In: Proceedings of the Genetic and Evolutionary Computation Conference, GECCO '18, pp. 466–473. Association for Computing Machinery, New York, NY, USA (2018). https://doi.org/10.1145/3205455.3205489.https://doi.org/10.1145/3205455.3205489

89. Suganuma, M., Shirakawa, S., Nagao, T.: A genetic programming approach to designing convolutional neural network architectures. In: Proceedings of the Genetic and Evolutionary Computation Conference, GECCO '17, pp. 497–504. Association for Computing Machinery, New York, NY, USA (2017). https://doi.org/10.1145/3071178.3071229, https://doi.org/10.1145/3071178.3071229

90. Sun, Y., Xue, B., Zhang, M., Yen, G.G.: Completely automated cnn architecture design based on blocks. IEEE Trans. Neural Netw. Learn. Syst. 31(4), 1242–1254 (2020). https://doi.org/10.1109/TNNLS.2019.2919608

91. Sun, Y., Xue, B., Zhang, M., Yen, G.G., Lv, J.: Automatically designing cnn architectures using the genetic algorithm for image classification. IEEE Trans. Cybern. 50(9), 3840–3854 (2020). https://doi.org/10.1109/TCYB.2020.2983860

92. Zhang, H., Jin, Y., Cheng, R., Hao, K.: Efficient evolutionary search of attention convolutional networks via sampled training and node inheritance. IEEE Trans. Evol. Comput. 25(2), 371–385 (2021)

93. Lu, Z., Whalen, I., Boddeti, V., Dhebar, Y., Deb, K., Goodman, E., Banzhaf, W.: Nsga-net: Neural architecture search using multi-objective genetic algorithm. In: Proceedings of the Genetic and Evolutionary Computation Conference, GECCO '19, pp. 419–427. Association for Computing Machinery, New York, NY, USA (2019). https://doi.org/10.1145/3321707.3321729, https://doi.org/10.1145/3321707.3321729

94. Jin, Y.: Surrogate-assisted evolutionary computation: Recent advances and future challenges. Swarm Evol. Comput. 1(2), 61–70 (2011)

95. Broomhead, D.S., Lowe, D.: Radial basis functions, multi-variable functional interpolation and adaptive networks. Technical report, Royal Signals and Radar Establishment Malvern (United Kingdom) (1988)

96. Dai, X., Zhang, P., Wu, B., Yin, H., Sun, F., Wang, Y., Dukhan, M., Hu, Y., Wu, Y., Jia, Y., et al.: Chamnet: towards efficient network design through platform-aware model adaptation. In: Proceedings of the IEEE/CVF Conference on Computer Vision and Pattern Recognition, pp. 11398–11407 (2019)

97. Jeong, S., Murayama, M., Yamamoto, K.: Efficient optimization design method using kriging model. J. Aircr. 42(2), 413–420 (2005)

98. Sun, Y., Wang, H., Xue, B., Jin, Y., Yen, G.G., Zhang, M.: Surrogate-assisted evolutionary deep learning using an end-to-end random forest-based performance predictor. IEEE Trans. Evol. Comput. 24(2), 350–364 (2020). https://doi.org/10.1109/TEVC.2019.2924461

99. Jin, Y., Wang, H., Sun, C.: Data-Driven Evolutionary Optimization. Springer, Berlin (2021)
100. Liu, S., Zhang, H., Jin, Y.: A survey on surrogate-assisted efficient neural architecture search. J. Autom. Learn. Syst. **1**(1) (2022)
101. Goldreich, O.: Secure multi-party computation. Manuscript. Preliminary version **78** (1998)
102. Dwork, C.: Differential privacy: a survey of results. In: Agrawal, M., Du, D., Duan, Z., Li, A. (eds.) Theory and Applications of Models of Computation, pp. 1–19. Springer, Berlin (2008)
103. Gentry, C.: A Fully Homomorphic Encryption Scheme. Stanford university (2009)
104. Yao, A.C.C.: How to generate and exchange secrets. In: 27th Annual Symposium on Foundations of Computer Science (sfcs 1986), pp. 162–167 (1986). https://doi.org/10.1109/SFCS.1986.25
105. Zhao, C., Zhao, S., Zhao, M., Chen, Z., Gao, C.Z., Li, H., Tan, Y.: Secure multi-party computation: theory, practice and applications. Inf. Sci. **476**, 357–372 (2019). https://doi.org/10.1016/j.ins.2018.10.024, https://www.sciencedirect.com/science/article/pii/S0020025518308338
106. Bellare, M., Desai, A., Jokipii, E., Rogaway, P.: A concrete security treatment of symmetric encryption. In: Proceedings 38th Annual Symposium on Foundations of Computer Science, pp. 394–403 (1997). https://doi.org/10.1109/SFCS.1997.646128
107. Rabin, M.O.: How to exchange secrets with oblivious transfer. Cryptology ePrint Archive, Report 2005/187 (2005). https://ia.cr/2005/187
108. Bonawitz, K., Ivanov, V., Kreuter, B., Marcedone, A., McMahan, H.B., Patel, S., Ramage, D., Segal, A., Seth, K.: Practical secure aggregation for privacy-preserving machine learning. In: Proceedings of the 2017 ACM SIGSAC Conference on Computer and Communications Security, CCS '17, pp. 1175–1191. Association for Computing Machinery, New York, NY, USA (2017). https://doi.org/10.1145/3133956.3133982, https://doi.org/10.1145/3133956.3133982
109. Keller, M.: MP-SPDZ: A Versatile Framework for Multi-Party Computation, pp. 1575–1590. Association for Computing Machinery, New York, NY, USA (2020). https://doi.org/10.1145/3372297.3417872
110. Damgård, I., Keller, M., Larraia, E., Pastro, V., Scholl, P., Smart, N.P.: Practical covertly secure mpc for dishonest majority - or: Breaking the spdz limits. In: Crampton, J., Jajodia, S., Mayes, K. (eds.) Computer Security—ESORICS 2013, pp. 1–18. Springer, Berlin (2013)
111. Damgård, I., Pastro, V., Smart, N., Zakarias, S.: Multiparty computation from somewhat homomorphic encryption. In: Safavi-Naini, R., Canetti, R. (eds.) Advances in Cryptology—CRYPTO 2012, pp. 643–662. Springer, Berlin (2012)
112. den Boer, B.: Diffie-hellman is as strong as discrete log for certain primes. In: Goldwasser, S. (ed.) Advances in Cryptology—CRYPTO' 88, pp. 530–539. Springer, New York (1990)
113. Dwork, C., McSherry, F., Nissim, K., Smith, A.: Calibrating noise to sensitivity in private data analysis. In: Theory of Cryptography Conference, pp. 265–284. Springer, Berlin (2006)
114. Dwork, C., Kenthapadi, K., McSherry, F., Mironov, I., Naor, M.: Our data, ourselves: Privacy via distributed noise generation. In: Annual International Conference on the Theory and Applications of Cryptographic Techniques, pp. 486–503. Springer, Berlin (2006)
115. Wei, K., Li, J., Ding, M., Ma, C., Yang, H.H., Farokhi, F., Jin, S., Quek, T.Q.S., Poor, H.V.: Federated learning with differential privacy: algorithms and performance analysis. IEEE Trans. Inf. Forensics Secur. **15**, 3454–3469 (2020). https://doi.org/10.1109/TIFS.2020.2988575
116. Abadi, M., Chu, A., Goodfellow, I., McMahan, H.B., Mironov, I., Talwar, K., Zhang, L.: Deep learning with differential privacy. In: Proceedings of the 2016 ACM SIGSAC Conference on Computer and Communications Security, CCS '16, pp. 308–318. Association for Computing Machinery, New York, (2016). https://doi.org/10.1145/2976749.2978318, https://doi.org/10.1145/2976749.2978318
117. Geyer, R.C., Klein, T., Nabi, M.: Differentially private federated learning: a client level perspective (2017). arXiv:1712.07557
118. Mahawaga Arachchige, P.C., Bertok, P., Khalil, I., Liu, D., Camtepe, S., Atiquzzaman, M.: Local differential privacy for deep learning. IEEE Internet Things J. **7**(7), 5827–5842 (2020). https://doi.org/10.1109/JIOT.2019.2952146

119. Zhao, Y., Zhao, J., Yang, M., Wang, T., Wang, N., Lyu, L., Niyato, D., Lam, K.Y.: Local differential privacy-based federated learning for internet of things. IEEE Internet Things J. **8**(11), 8836–8853 (2021). https://doi.org/10.1109/JIOT.2020.3037194

120. Seif, M., Tandon, R., Li, M.: Wireless federated learning with local differential privacy. In: 2020 IEEE International Symposium on Information Theory (ISIT), pp. 2604–2609 (2020). https://doi.org/10.1109/ISIT44484.2020.9174426

121. Truex, S., Liu, L., Chow, K.H., Gursoy, M.E., Wei, W.: Ldp-fed: Federated learning with local differential privacy. In: Proceedings of the Third ACM International Workshop on Edge Systems, Analytics and Networking, EdgeSys '20, pp. 61–66. Association for Computing Machinery, New York (2020). https://doi.org/10.1145/3378679.3394533, https://doi.org/10.1145/3378679.3394533

122. Rivest, R.L., Shamir, A., Adleman, L.M.: A method for obtaining digital signatures and public key cryptosystems. In: Secure Communications and Asymmetric Cryptosystems, pp. 217–239. Routledge (2019)

123. Montgomery, P.L.: A survey of modern integer factorization algorithms. CWI Q. **7**(4), 337–366 (1994)

124. Paillier, P.: Public-key cryptosystems based on composite degree residuosity classes. In: Stern, J. (ed.) Advances in Cryptology—EUROCRYPT '99, pp. 223–238. Springer, Berlin (1999)

125. Shamir, A.: How to share a secret. Commun. ACM **22**(11), 612–613 (1979). https://doi.org/10.1145/359168.359176

126. Zhang, C., Li, S., Xia, J., Wang, W., Yan, F., Liu, Y.: BatchCrypt: Efficient homomorphic encryption for Cross-Silo federated learning. In: 2020 USENIX Annual Technical Conference (USENIX ATC 20), pp. 493–506. USENIX Association (2020). https://www.usenix.org/conference/atc20/presentation/zhang-chengliang

127. Lyubashevsky, V.: Lattice-based identification schemes secure under active attacks. In: Cramer, R. (ed.) Public Key Cryptography—PKC 2008, pp. 162–179. Springer, Berlin (2008)

128. van Dijk, M., Gentry, C., Halevi, S., Vaikuntanathan, V.: Fully homomorphic encryption over the integers. In: Gilbert, H. (ed.) Advances in Cryptology—EUROCRYPT 2010, pp. 24–43. Springer, Berlin (2010)

129. Levieil, E., Naccache, D.: Cryptographic test correction. In: International Workshop on Public Key Cryptography, pp. 85–100. Springer, Berlin (2008)

130. Cheon, J.H., Kim, A., Kim, M., Song, Y.: Homomorphic encryption for arithmetic of approximate numbers. In: Takagi, T., Peyrin, T. (eds.) Advances in Cryptology—ASIACRYPT 2017, pp. 409–437. Springer International Publishing, Cham (2017)

131. Yagisawa, M.: Fully homomorphic encryption without bootstrapping. Cryptology ePrint Archive, Report 2015/474 (2015). https://ia.cr/2015/474

132. Ma, J., Naas, S.A., Sigg, S., Lyu, X.: Privacy-preserving federated learning based on multi-key homomorphic encryption. Int. J. Intell. Syst. (2022)

133. Jiang, Z., Wang, W., Liu, Y.: Flashe: Additively symmetric homomorphic encryption for cross-silo federated learning (2021). arXiv:2109.00675

134. Yang, Q., Liu, Y., Chen, T., Tong, Y.: Federated machine learning: concept and applications. ACM Trans. Intell. Syst. Technol. (TIST) **10**(2), 1–19 (2019)

135. Haddadpour, F., Mahdavi, M.: On the convergence of local descent methods in federated learning (2019). arXiv:1910.14425

136. Liu, Y., Kang, Y., Xing, C., Chen, T., Yang, Q.: A secure federated transfer learning framework. IEEE Intell. Syst. **35**(4), 70–82 (2020). https://doi.org/10.1109/MIS.2020.2988525

137. Yu, F., Zhang, W., Qin, Z., Xu, Z., Wang, D., Liu, C., Tian, Z., Chen, X.: Heterogeneous federated learning (2020). arXiv:2008.06767

138. Aledhari, M., Razzak, R., Parizi, R.M., Saeed, F.: Federated learning: a survey on enabling technologies, protocols, and applications. IEEE Access **8**, 140699–140725 (2020). https://doi.org/10.1109/ACCESS.2020.3013541

139. Cheng, K., Fan, T., Jin, Y., Liu, Y., Chen, T., Papadopoulos, D., Yang, Q.: SecureBoost: a lossless federated learning framework. IEEE Intell. Syst. **86**(6), 87–98 (2021)

140. Zhu, H., Xu, J., Liu, S., Jin, Y.: Federated learning on non-iid data: a survey. Neurocomputing **465**, 371–390 (2021). https://doi.org/10.1016/j.neucom.2021.07.098.
141. McMahan, H.B., Moore, E., Ramage, D., y Arcas, B.A.: Federated learning of deep networks using model averaging (2016). arXiv:1602.05629 (2016)
142. Yao, X., Huang, T., Wu, C., Zhang, R.X., Sun, L.: Federated learning with additional mechanisms on clients to reduce communication costs (2019). arXiv:1908.05891
143. Xu, J., Jin, Y., Du, W., Gu, S.: A federated data-driven evolutionary algorithm. Knowl.-Based Syst. **233**, 107532 (2021)
144. Pan, S.J., Yang, Q.: A survey on transfer learning. IEEE Trans. Knowl. Data Eng. **22**(10), 1345–1359 (2010). https://doi.org/10.1109/TKDE.2009.191
145. Su, H., Maji, S., Kalogerakis, E., Learned-Miller, E.G.: Multi-view convolutional neural networks for 3d shape recognition. In: 2015 IEEE International Conference on Computer Vision, ICCV 2015, pp. 945–953. IEEE Computer Society, Santiago, Chile (2015). https://doi.org/10.1109/ICCV.2015.114
146. Li, Q., Diao, Y., Chen, Q., He, B.: Federated learning on non-iid data silos: an experimental study (2021). arXiv:2102.02079
147. McMahan, B., Moore, E., Ramage, D., Hampson, S., Arcas, B.A.y.: Communication-Efficient Learning of Deep Networks from Decentralized Data. In: Singh, A., Zhu, J. (eds.) Proceedings of the 20th International Conference on Artificial Intelligence and Statistics, Proceedings of Machine Learning Research, vol. 54, pp. 1273–1282. PMLR (2017). https://proceedings.mlr.press/v54/mcmahan17a.html
148. Yurochkin, M., Agarwal, M., Ghosh, S., Greenewald, K.H., Hoang, T.N., Khazaeni, Y.: Bayesian nonparametric federated learning of neural networks. In: Proceedings of the 36th International Conference on Machine Learning, ICML 2019, Long Beach, California, USA, Proceedings of Machine Learning Research, vol. 97, pp. 7252–7261. PMLR (2019)
149. Li, Q., He, B., Song, D.: Model-agnostic round-optimal federated learning via knowledge transfer (2020). arXiv:2010.01017
150. Lin, T., Kong, L., Stich, S.U., Jaggi, M.: Ensemble distillation for robust model fusion in federated learning. In: 34th Conference on Neural Information Processing Systems (NeurIPS 2020) (2020)
151. Wang, H., Yurochkin, M., Sun, Y., Papailiopoulos, D., Khazaeni, Y.: Federated learning with matched averaging. In: International Conference on Learning Representations (2020). https://openreview.net/forum?id=BkluqlSFDS
152. Wang, J., Liu, Q., Liang, H., Joshi, G., Poor, H.V.: Tackling the objective inconsistency problem in heterogeneous federated optimization. In: Advances in Neural Information Processing Systems, vol. 33, pp. 7611–7623. Curran Associates, Inc. (2020)
153. Garcia-Molina, H., Joglekar, M., Marcus, A., Parameswaran, A., Verroios, V.: Challenges in data crowdsourcing. IEEE Trans. Knowl. Data Eng. **28**(4), 901–911 (2016)
154. Zhao, Y., Li, M., Lai, L., Suda, N., Civin, D., Chandra, V.: Federated learning with non-iid data (2018). arXiv:1806.00582, https://arxiv.org/pdf/1806.00582.pdf
155. Tuor, T., Wang, S., Ko, B.J., Liu, C., Leung, K.K.: Overcoming noisy and irrelevant data in federated learning. In: 2020 25th International Conference on Pattern Recognition (ICPR), pp. 5020–5027 (2021).https://doi.org/10.1109/ICPR48806.2021.9412599
156. Yoshida, N., Nishio, T., Morikura, M., Yamamoto, K., Yonetani, R.: Hybrid-fl for wireless networks: cooperative learning mechanism using non-iid data. In: ICC 2020—2020 IEEE International Conference on Communications (ICC), pp. 1–7 (2020). https://doi.org/10.1109/ICC40277.2020.9149323
157. Tanner, M.A., Wong, W.H.: The calculation of posterior distributions by data augmentation. J. Am. Stat. Assoc. **82**(398), 528–540 (1987)
158. Duan, M., Liu, D., Chen, X., Tan, Y., Ren, J., Qiao, L., Liang, L.: Astraea: self-balancing federated learning for improving classification accuracy of mobile deep learning applications. In: 2019 IEEE 37th International Conference on Computer Design (ICCD), pp. 246–254. IEEE (2019)

159. Zhang, H., Cissé, M., Dauphin, Y.N., Lopez-Paz, D.: mixup: beyond empirical risk minimization. In: 6th International Conference on Learning Representations, ICLR 2018, Vancouver, BC, Canada, Conference Track Proceedings. OpenReview.net (2018)

160. Goodfellow, I., Pouget-Abadie, J., Mirza, M., Xu, B., Warde-Farley, D., Ozair, S., Courville, A., Bengio, Y.: Generative adversarial nets. In: Ghahramani, Z., Welling, M., Cortes, C., Lawrence, N., Weinberger, K.Q. (eds.) Advances in Neural Information Processing Systems, vol. 27. Curran Associates, Inc. (2014). https://proceedings.neurips.cc/paper/2014/file/5ca3e9b122f61f8f06494c97b1afccf3-Paper.pdf

161. Shin, M., Hwang, C., Kim, J., Park, J., Bennis, M., Kim, S.L.: Xor mixup: Privacy-preserving data augmentation for one-shot federated learning (2020). arXiv:2006.05148

162. Li, T., Sahu, A.K., Zaheer, M., Sanjabi, M., Talwalkar, A., Smith, V.: Federated optimization in heterogeneous networks. In: Dhillon, I., Papailiopoulos, D., Sze, V. (eds.) Proceedings of Machine Learning and Systems, vol. 2, pp. 429–450 (2020). https://proceedings.mlsys.org/paper/2020/file/38af86134b65d0f10fe33d30dd76442e-Paper.pdf

163. Arivazhagan, M.G., Aggarwal, V., Singh, A.K., Choudhary, S.: Federated learning with personalization layers (2019). arXiv:1912.00818

164. Smith, V., Chiang, C., Sanjabi, M., Talwalkar, A.S.: Federated multi-task learning. In: Advances in Neural Information Processing Systems 30: Annual Conference on Neural Information Processing Systems 2017, pp. 4424–4434. Long Beach, CA, USA (2017)

165. Jaggi, M., Smith, V., Takac, M., Terhorst, J., Krishnan, S., Hofmann, T., Jordan, M.I.: Communication-efficient distributed dual coordinate ascent. In: Ghahramani, Z., Welling, M., Cortes, C., Lawrence, N., Weinberger, K. (eds.) Advances in Neural Information Processing Systems, vol. 27. Curran Associates, Inc. (2014). https://proceedings.neurips.cc/paper/2014/file/894b77f805bd94d292574c38c5d628d5-Paper.pdf

166. Liu, S., Pan, S.J., Ho, Q.: Distributed multi-task relationship learning. In: Proceedings of the 23rd ACM SIGKDD International Conference on Knowledge Discovery and Data Mining, KDD '17, pp. 937–946. Association for Computing Machinery, New York, NY, USA (2017). https://doi.org/10.1145/3097983.3098136, https://doi.org/10.1145/3097983.3098136

167. Ma, C., Smith, V., Jaggi, M., Jordan, M., Richtarik, P., Takac, M.: Adding vs. averaging in distributed primal-dual optimization. In: Bach, F., Blei, D. (eds.) Proceedings of the 32nd International Conference on Machine Learning, Proceedings of Machine Learning Research, vol. 37, pp. 1973–1982. PMLR, Lille, France (2015). https://proceedings.mlr.press/v37/mab15.html

168. Hinton, G., Vinyals, O., Dean, J.: Distilling the knowledge in a neural network (2015). arXiv:1503.02531

169. Lin, T., Kong, L., Stich, S.U., Jaggi, M.: Ensemble distillation for robust model fusion in federated learning. In: Larochelle, H., Ranzato, M., Hadsell, R., Balcan, M., Lin, H. (eds.) Advances in Neural Information Processing Systems, vol. 33, pp. 2351–2363. Curran Associates, Inc. (2020). https://proceedings.neurips.cc/paper/2020/file/18df51b97ccd68128e994804f3eccc87-Paper.pdf

170. Duchi, J.C., Hazan, E., Singer, Y.: Adaptive subgradient methods for online learning and stochastic optimization. In: COLT 2010—The 23rd Conference on Learning Theory, pp. 257–269. Haifa, Israel, Omnipress (2010)

171. Kingma, D.P., Ba, J.: Adam: A method for stochastic optimization. In: 3rd International Conference on Learning Representations, ICLR 2015, Conference Track Proceedings. San Diego, CA, USA (2015)

172. Ioffe, S., Szegedy, C.: Batch normalization: accelerating deep network training by reducing internal covariate shift. In: Bach, F., Blei, D. (eds.) Proceedings of the 32nd International Conference on Machine Learning, Proceedings of Machine Learning Research, vol. 37, pp. 448–456. PMLR, Lille, France (2015). https://proceedings.mlr.press/v37/ioffe15.html

173. Mansour, Y., Mohri, M., Ro, J., Suresh, A.T.: Three approaches for personalization with applications to federated learning (2020). arXiv:2002.10619

174. Kopparapu, K., Lin, E.: Fedfmc: Sequential efficient federated learning on non-iid data (2020). arXiv:2006.10937

175. Ghosh, A., Hong, J., Yin, D., Ramchandran, K.: Robust federated learning in a heterogeneous environment (2019). arXiv:1906.06629
176. Ghosh, A., Chung, J., Yin, D., Ramchandran, K.: An efficient framework for clustered federated learning. In: Advances in Neural Information Processing Systems 33: Annual Conference on Neural Information Processing Systems 2020, NeurIPS 2020, virtual (2020)
177. Khan, L.U., Pandey, S.R., Tran, N.H., Saad, W., Han, Z., Nguyen, M.N.H., Hong, C.S.: Federated learning for edge networks: resource optimization and incentive mechanism. IEEE Commun. Mag. **58**(10), 88–93 (2020). https://doi.org/10.1109/MCOM.001.1900649

Chapter 2
Communication Efficient Federated Learning

Abstract Federated learning, as a branch of distributed learning, suffers from the high cost and concurrency of communication. The situation gets even worse in training deep models on large-scale mobile devices. Hence, communication efficiency is an avoidable topic in multi-party federated learning systems. Focusing on this issue, this chapter first discusses the background, the main methodologies and potential directions for communication efficient federated learning. Then, we introduce two methods for reducing the communication cost of federated learning, which are layer-wise asynchronous update and model quantization. Extensive experiments are performed to show the efficiency of these two approaches.

2.1 Communication Cost in Federated Learning

Recall the definition of FedAvg, the global model is generated by aggregating the local models based on the weighted sum approach:

$$w_t = \sum_{k=1}^{\lambda K} p_k w_t^k, \tag{2.1}$$

To achieve this, all participating clients must communicate with the server twice in each round of model update, uploading the trained local model parameters and downloading the averaged global model. For example, given a vanilla federated learning system with 100 clients and 20% participation ratio, a deep network with 30 Mb model size is co-trained. Then, in each communication round the sum of the amount of data uploaded and downloaded is about 1.2 Gb, and usually the communication is high concurrency. Besides, the average upstream and downstream bandwidths are asymmetric, e.g., the mean mobile download speed is 26.36 Mbps while the upload speed is 11.05 Mbps in Q3-Q4 2017 in the UK [1].

Therefore, communication efficiency is an important topic in the context of federated learning [2]. And recently, a number of methods have been developed to reduce the communication cost in federated learning, including but are not limited to model compression, incremental update, asynchronous averaging, and neural architecture search (NAS). Among them, the starting point of asynchronous aggregation is the barrel effect of client computing power, which is from the perspective of practical problems. Besides, model compression is also one of the mainstream methods, since it can not only reduce the communication cost, but also reduce the inference time of the clients. And hence, in this chapter, we will introduce these two types of approaches and their applications in reducing communication cost in federated learning.

In Sect. 2.3, a layer-wise asynchronous aggregation method is presented, which is a direct application of asynchronous methods. And in Sect. 2.4, a quantization-based federated model compression method is introduced.

2.2   Main Methodologies

Reducing communication cost is of great importance in federated learning research and its practical applications in various scenarios, for which communication-efficient strategies lie in the core. This encompasses a myriad of valuable explorations, including providing compression techniques for federated learning, making wireless-FL co-design, replacing the exchanged parameters with light-weight knowledge via employing knowledge distillation, and designing optimization algorithms with higher efficiency [3].

What is worse, the major fundamental challenges in federated learning, including non-IID data distribution and unbalanced data, and massively distributed participating clients [4] also pose challenges to improving communication efficiency. In the following, we begin with briefly surveying these challenges together with the relevant concepts.

2.2.1   Non-IID/IID Data and Dataset Shift

Local data are user-generated contents and heavily rely on the specific users, presenting strong non-identicalness. From the perspective of the data distribution differences on different clients, the taxonomy of Non-IID/IID data is closely related to dataset shift. In addition, the temporal dependence of the distributions of either the participating clients or their corresponding local data possibly also introduces dataset shift in the classic sense, including differences between data distributions in the training and test data, and differences between the development and deployment distributions [5–7].

2.2.2 Non-identical Client Distributions

Kairouz et al. [5] summarize the non-identical client distributions in FL into the following categories. (1) Feature Distribution Skew (covariate shift), (2) Label Distribution Skew (prior probability shift), (3) Same Label, Different Features (concept drift), (4) Same Features, Different Label (concept shift), and (5) Quantity Skew or Unbalancedness. Datasets in FL in the real world may present a mixture of these categories. Most FL studies consider label distribution skew by partitioning existing IID datasets to form the corresponding category based on the labels.

2.2.3 Violations of Independence

The over-training of the local models changes the distribution of the participating clients leads to violations of independence. For instance, cross-device FL involving many different types of devices introduces a diurnal-pattern bias in data sources. In addition, device locations, especially the change in longitude, introduce a huge geographic bias. Eichner et al. [8] discussed this problem and a few mitigation strategies, but lots of challenges remain.

Studies on the combinations of FedAVG with the sparsification or quantization of update gradients have significantly reduced the communication cost with only minor accuracy degradation [9]. However, further reduction of communication remains challenging, as there is a trade-off between accuracy and communication efficiency. Such fundamental trade-offs have been approached from the theoretical statistics perspective [10–13]; however, it remains open as how to leverage such statistical approaches to enhance practical training methods. According to the chosen compression objectives, these approaches can be categorised into gradient compression [9, 12, 14], model broadcast compression [15], and local computation reduction [16, 17].

In the specific scenarios relying on wireless communication, the wireless-FL co-design may bring benefits for reducing training latency and thus enhance the reliability of the entire production system by fully making use of the wireless channel dynamics during training. Wireless interference, including noisy channels and channel fluctuations, may degrade communications between the server and clients, causing systemic latency and deteriorating the reliability. A few studies optimize the clients' scheduling policy with regard to computing and communication resources [18, 19]. In addition, Abari et al. develop natural data aggregators by leveraging the unique characteristics of wireless channels, e.g., broadcast and superposition [20].

Federated Distillation (FD) aims to reduce the data exchanged between the server and clients by making use of knowledge distillation techniques, where the exchanged are local-model outputs (termed as logits) rather than model parameters [21, 22].

Thus, federated learning becomes lightweight and practical. The rest of this section focuses on the design of federated learning with better communication efficiency and compression techniques.

2.3 Temporally Weighted Averaging and Layer-Wise Weight Update

This section introduces the design of federated learning with a layer-wise asynchronous model update and temporally weighted aggregation, called ASTW _FedAVG. The main motivation of the layer-wise asynchronous model update is inspired from the observation that the shallow layers (those closer to the input layer) need to be updated more frequently than the deep layers. Consequently, only the shallow layers need to be updated in each round, while the deep layers will be communicated to the server less frequently. Meanwhile, since the local models are updated in different rounds, the weighted aggregation of the local models should take the timeliness of different local models into account, which is called temporally weighted aggregation. The overall framework is illustrates in Fig. 2.1 and the pseudo-code run on the server and clients is listed in Algorithms 1 and 2, respectively.

The code for the server consists of an initialization step followed by a number of communication rounds. In initialization (Algorithm 1, Lines 2–6), the central model ω_0, timestamps $timestamp_g$ and $timestamp_s$ are initialized. Timestamps are stored and used to weight the timeliness of the corresponding parameters in aggregation.

The training process is divided into three steps. Lines 8-12 in Algorithm 1 set $flag$ to be true in the last $1/freq$ rounds in each loop. Assume there are $rounds\_in\_loop$ rounds in each loop. Lines 13–14 randomly select a participating subset S_t of the clients according to C, which is the fraction of participating clients per round. In

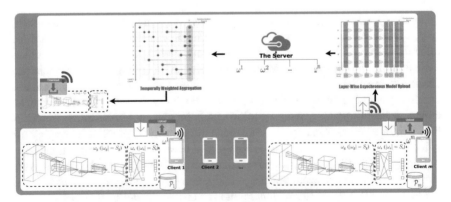

Fig. 2.1 Federated learning with layer-wise asynchronous model update and temporally weighted aggregation

---

Algorithm 1 Server Component of ASTW_FedAVG

1: **function** SERVEREXECUTION ▷ Run on the server
2:    initialize ω_0
3:    **for** each client $k \in \{1, 2, ..., K\}$ **do**
4:        $timestamp_g^k \leftarrow 0$
5:        $timestamp_s^k \leftarrow 0$
6:    **end for**
7:    **for** each round $t = 1, 2, ...$ **do**
8:        **if** t mod $rounds\_in\_loop \in set_{ES}$ **then**§
9:            $flag \leftarrow$ True
10:        **else**
11:            $flag \leftarrow$ False
12:        **end if**
13:        $m \leftarrow \max(C * K, 1)$
14:        $S_t \leftarrow$ (random set of m clients)
15:        **for** each client $k \in S_t$ **in parallel do**
16:            **if** $flag$ **then**
17:                $\omega^k \leftarrow$ ClientUpdate$(k, \omega_t, flag)$
18:                $timestamp_g^k \leftarrow t$
19:                $timestamp_s^k \leftarrow t$
20:            **else**
21:                $\omega_g^k \leftarrow$ ClientUpdate$(k, \omega_{g,t}, flag)$
22:                $timestamp_g^k \leftarrow t$
23:            **end if**
24:        **end for**
25:        $\omega_{g,t+1} \leftarrow \sum_{k=1}^{K} \frac{n_k}{n} * f_g(t,k) * \omega_g^k$ †
26:        **if** $flag$ **then**
27:            $\omega_{s,t+1} \leftarrow \sum_{k=1}^{K} \frac{n_k}{n} * f_s(t,k) * \omega_s^k$ ‡
28:        **end if**
29:    **end for**
30: **end function**
    § $rounds\_in\_loop = 15$ and $set_{ES} = \{11, 12, 13, 14, 0\}$
    † $f_g(t,k) = a^{-(t-timestamp_g^k)}$
    ‡ $f_s(t,k) = a^{-(t-timestamp_s^k)}$

---

lines 15–24, sub-function *ClientUpdate* is called in parallel to get ω^k/ω_g^k, and the corresponding timestamps are updated. $flag$ specifies whether all layers or only the shallow layers will be updated and communicated. Then in lines 25–28, the aggregation is performed to update ω_g. Note that compared with Eq. 2.2, a parameter a is introduced into the weighting function in line 25 or line 27 to examine the influence of different weightings in the experiments. In this work, a is set to e or $e/2$.

The implementation of the local model update (Algorithm 2) is controlled by three parameters, k, ω, and $flag$, where k is the index of the selected client, $flag$ indicates whether all layers or only the shallow layers will be updated. B and E denote the local mini-batch size and the local epoch, respectively. In Algorithm 2, Line 2 splits data into batches, whereas Lines 3–7 set all layers or shallow layers of the local model to be downloaded according to $flag$. In lines 8–12, the local SGD is performed. Lines 13–17 return the local parameters.

Algorithm 2 Client Component of ASTW_FedAVG

```
1: function CLIENTUPDATE(k, w, flag)                              ▷ Run on client k
2:     B ← (split P_k into batches of size B)
3:     if flag then
4:         ω ← ω
5:     else
6:         ω_s ← ω
7:     end if
8:     for each local epoch i from 1 to E do
9:         for batch b ∈ B do
10:            ω ← ω − η * ∇ℓ(w; b)
11:        end for
12:    end for
13:    if flag then
14:        return ω to server
15:    else
16:        return ω_s to server
17:    end if
18: end function
```

The proposed approach consists of two main components, as suggested by its name. A detailed description of the main components will be given in the following subsection.

2.3.1 Temporally Weighted Averaging

When conducting aggregation, previous mainstream federated learning focuses on the data amount of each client. In the example illustrated in Fig. 2.2, a local model from client k with more data (the larger n_k) will have the stronger influence on the global model. In the figure, the **blue diamonds** represent the out-of-date checkpoints of the local models that do not take part in the $t + 1$th update of the central model and the **orange dots** represent up-to-date checkpoints of the local models that participate in the current aggregation. As shown in Algorithm 3 (Line 9), the aggregation strategy weights those participating local models (orange dots in Fig. 2.2) only by their data size regardless the timing of local model update. In other words, the local models updated in round $t − p$ and these updated in round t are equally weighted, no matter how large p is. This weighting approach may be not reasonable since the weights of different models are updated in different rounds.

However, federated learning deals with incremental and streaming data that are continuously generated, discarded, and distributed on different sources, i.e., participating clients. Thus, the aggregation should attach a larger weight to newer checkpoints. Accordingly, the following model aggregation method takes into account the timestamp when the local model is updated.

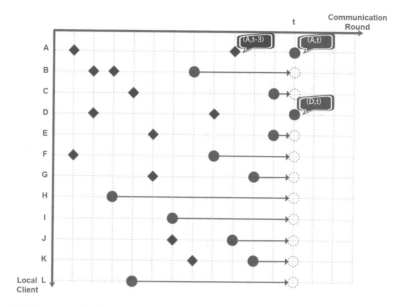

Fig. 2.2 The conventional aggregation strategy

$$\omega_{t+1} \leftarrow \sum_{k=1}^{K} \frac{n_k}{n} * (e/2)^{-(t-timestamp^k)} * \omega^k \qquad (2.2)$$

where e is the natural logarithm used to depict the time effect, t means the current round, and $timestamp^k$ is the round in which the newest ω^k was updated. Fig. 2.3 illustrates the proposed temporally weighted aggregation, where, similar in Fig. 2.2, the orange dots denote the participating clients and the blue diamonds represent the rest. In the figure, the degree of transparency is used to indicate weight of the corresponding local model in the aggregation.

2.3.2 Layer-Wise Asynchronous Weight Update

The most intrinsic requirement for reducing the communication cost in federated learning is to upload/download as little data as possible without deteriorating the performance of the central model. To this end, this section presents a layer-wise asynchronous model update strategy that updates the local model parameters of different layers at different frequencies to reduce the amount of data to be exchanged. This idea was primarily inspired by the following interesting observations made in fine-tuning deep neural networks [23, 24]:

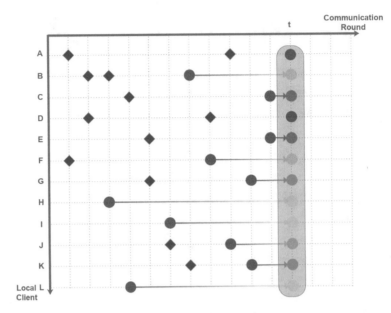

Fig. 2.3 An illustration of the temporally weighted aggregation

---

Algorithm 3 FedAVG [4].

1: **function** SERVEREXECUTION   ▷ Run on the server
2:     initialize w_0
3:     **for** each round $t = 1, 2, ...$ **do**
4:         $m \leftarrow \max(C \cdot K, 1)$
5:         $S_t \leftarrow$ (random set of m clients)
6:         **for** each client $k \in S_t$ **in parallel do**
7:             $w_{t+1}^k \leftarrow$ ClientUpdate(k, w_t)
8:         **end for**
9:         $w_{t+1} \leftarrow \sum_{k=1}^{K} \frac{n_k}{n} w_{t+1}^k$
10:    **end for**
11: **end function**
12:
13: **function** CLIENTUPDATE(k, w) ▷ Run on client k
14:     $\mathcal{B} \leftarrow$ (split $\mathcal{P}_k$ into batches of size B)
15:     **for** each local epoch i from 1 to E **do**
16:         **for** batch $b \in \mathcal{B}$ **do**
17:             $w \leftarrow w - \eta \bigtriangledown \ell(w; b))$
18:         **end for**
19:     **end for**
20:     return w to server
21: **end function**
---

1. Shallow layers corresponding to general features make contributions learning features across different tasks/datasets, although they are a relatively small part of the deep neural network model in terms of the number of parameters.

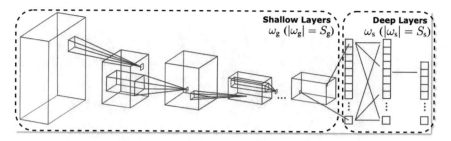

Fig. 2.4 An illustration of the shallow and deep layers of a deep neural network

2. By contrast, deep layers learn ad hoc features that are tasks or datasets specific. The majority of a deep learning model belong to the deep layers.

Given the above observations, the shallow layers with a relatively smaller number of parameters are more pivotal for federated learning, considering their contributions to the aggregated central model. Accordingly, the update frequency of shallow layers should be higher than that of deep ones. Therefore, the parameters in the shallow and deep layers can be updated asynchronously, thereby reducing the amount of communication between the server and clients. We term this layer-wise asynchronous model update.

Figure 2.4 shows a typical deep neural network (DNN) with shallow and deep layers, and the local clients under consideration adopt a DNN as the model architecture. Concretely, the shallow layers extracting general features are denoted as ω_g, and the deep layers learning specific features are denoted as ω_s. Their sizes are S_g and S_s, respectively, and typically, $S_g \ll S_s$. Compared to ω_s, the proposed asynchronous learning strategy updates and exchanges ω_g at a higher frequency. Federated learning divides its whole process into loops, and each one having T rounds of model updates. In a loop, the server updates ω_g and exchanges it with clients in all rounds but only conducts these operations on ω_s in only fe rounds, where $fe < T$. Therefore, the improvement reduces communication cost, in terms of a reduced number of parameters to be transmitted, i.e., $fe * S_s$. This way, the communication cost can be significantly reduced, since S_s is usually very large in DNNs.

Figure 2.5 illustrates the asynchronous learning strategy with an example. The abscissa corresponds to the communication round, and the ordinate denotes the local client. The discussed example involves a server and five local devices, $\{A, B, C, D, E\}$. In round t, client A participating in the global model update corresponds to point (A, t).

As shown in Fig. 2.5(a), the server in a conventional synchronous aggregation strategy uploads/downloads both ω_g and ω_s in all rounds. The illustration uses grey rounded rectangles in the bottom to represent the aggregation of the shallow and deep layers. By contrast, Fig. 2.5(b) shows the proposed asynchronous learning strategy. The example provides a loop of six rounds $(t-5, t-4, \ldots, t)$, and only rounds $t-1$ and t exchange deep-layer parameters. Hence, the number of reduced parameters to be communicated is $2/3 * S_s$.

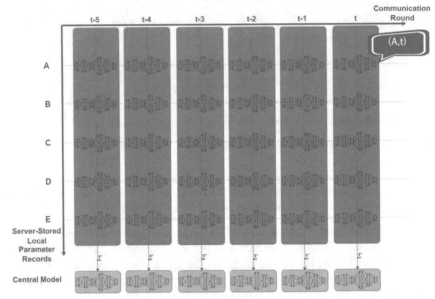

(a) Synchronous model update strategy.

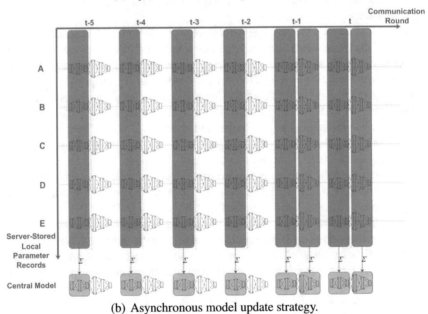

(b) Asynchronous model update strategy.

Fig. 2.5 Parameter exchange

2.3.3 Empirical Studies

Experimental Settings With the purpose to evaluate the proposed ASTW_FedAVG in terms of the model performance and communication cost, empirical studies involving (1) convolution neural networks (CNNs) for image processing and (2) long short-term memory (LSTM) for human activity recognition are conducted. Specifically, the CNN model adopts the structure of two stacked convolution layers and one fully connected layer [25, 26] to address the MNIST handwritten digit recognition task, and the LSTM consists of two stacked LSTM layers and one fully connected layer to accomplish the human activity recognition task [27, 28]. Data split is executed on the two datasets so that the evaluation can fully reflect those different real-world scenarios where the proposed federated learning framework is supposed to be deployed, e.g., non-IID distribution, unbalanced amount, and massively decentralized datasets. More details can be found in the subsequent section.

The empirical study selects the state-of-the-art federated averaging (FedAVG, detailed in Algorithm 3) [4] as the baseline algorithm. Comparisons are made among four algorithms, i.e., the baseline FedAVG, the proposed ASTW_FedAVG, TW_FedAVG, and AS_FedAVG. The last two are variants of the proposed algorithm, namely, TWFL that adopts temporally weighted aggregation without the layer-wise asynchronous model update, and AS_FedAVG that employs the layer-wise asynchronous model update without using temporally weighted aggregation.

Table 2.1 lists the most significant parameters of the proposed ASTW_FedAVG. $freq$ controls the update/exchange frequency of deep-layer parameters ω_s per loop. For instance, $freq = 5/15$ means that only in the last five of the 15 rounds, the server and clients update and exchange ω_s. In model aggregation, a adjusts the time effect. K and m are environmental parameters controlling the scale or complexity of the experiments, K denotes the number of local clients, and m is the number of participating clients per round.

Settings on Datasets
As previously discussed, the particular challenges, such as non-IID, imbalanced, and massively decentralized datasets, characterize federated learning. Therefore, the empirical study should adopt a data split method that reflects these challenges. Algorithm 4 details the method for generating the client datasets, in which $Labels$, N_c, S_{min}, and S_{max} are three parameters, denoting the names of classes involved in

Table 2.1 Parameter settings

Notion	Parameter range
$freq$	$\{3/15, 5/15^*, 7/15\}$
a	$\{e/2^*, e\}$
K	$\{10, 20^*\}$
m	$\{1, 2^*\}$

*Default setting

the corresponding tasks, the number of classes in each local dataset, and the lower and upper bounds of the size of the local data.

Algorithm 4 Generation of Local Datasets.

Input: $Labels$, N_C, S_{min}, and S_{max}
Output: Non-IID and unbalanced local dataset $\mathcal{P}_k$
1: $classes \leftarrow$ Choices($Labels, N_C$)
2: $L \leftarrow$ Len($Labels$)
3: $weights \leftarrow$ Zeros(L)
4: $\mathcal{P} \leftarrow$ Zeros(L)
5: **for** each $class \in classes$ **do**
6:    $weights_{class} \leftarrow$ Random(0,1)
7: **end for**
8: $sum \leftarrow \sum_{class=1}^{L} weights_{class}$
9: $num \leftarrow$ Random(S_{min}, S_{max})
10: **for** each $class \in classes$ **do**
11:    $\mathcal{P}_{class} \leftarrow \frac{weights_{class}}{sum} \times num$
12: **end for**
13: $\mathcal{P}_k \leftarrow \mathcal{P}$

Handwritten Digit Recognition Using CNN MNIST is a dataset of handwritten digit images, each corresponding to a digit of 0–9, and all are 28×28-pixel gray-scale. When partitioning data over local clients, the data split scheme first sorts them by their labels, and gets ten shards. Then, Algorithm 4 computes $\mathcal{P}_k$, where k indexes clients. To each client, $\mathcal{P}_k$ indicates the partition coefficients corresponding to these shards. In this task, $Labels = \{0, 1, 2, \ldots, 9\}$; N_c is randomly chosen from $\{2, 3\}$, given $K = 20$, $S_{min} = 1000$, and $S_{max} = 1600$. For easy analyses, five partitions/local datasets, namely 1@MNIST, 2@MNIST, ..., 5@MNIST are predefined. Figure 2.6 plots their corresponding 3-D column.

Table 2.2 lists hyper-parameters of the adopted CNN, which stacks two 5×5 convolution layers (a 32-channel one and a 64-channel one) and 2×2 max-pooling layers. Then, a fully connected layer of 512 units with a ReLU activation function is attached. The output layer is a softmax unit.

Table 2.2 Parameters settings for the CNN

Layer	Shapes
conv2d_1	$5 \times 5 \times 1 \times 32$
conv2d_1	32
conv2d_2	$5 \times 5 \times 32 \times 64$
conv2d_2	64
dense_1	1024×512
dense_1	512
dense_2	512×10
dense_2	10

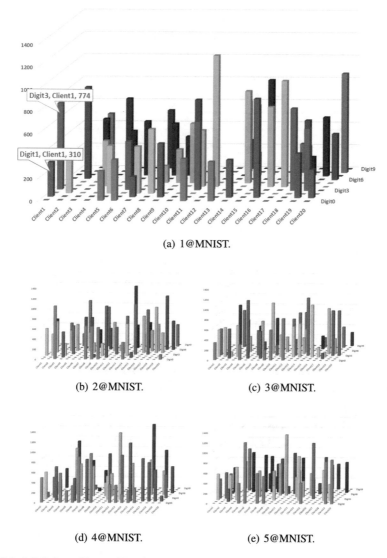

(a) 1@MNIST.

(b) 2@MNIST.

(c) 3@MNIST.

(d) 4@MNIST.

(e) 5@MNIST.

Fig. 2.6 3-D Column Charts of Pre-Generated Datasets

Human Activity Recognition using LSTM

HAR is a dataset of sensor-generated time series with a label out of six activities. The study applies a similar split operation on HAR to distribute data over local clients. Here, $Labels = \{0, 1, 2, \ldots, 5\}$; N_c is randomly chosen from $\{2, 3\}$, given $K = 20$, $S_{min} = 250$, and $S_{max} = 500$.

Table 2.3 details the LSTM hyper-parameters. The architecture adopted by this study stacks two 5×5 LSTM layers (the first one with $cell\_size = 20$ and

Table 2.3 Parameter settings for the LSTM

Layer	Shapes
lstm_1	9×100
lstm_1	25×100
lstm_1	100
lstm_2	25×100
lstm_2	25×100
lstm_2	100
dense_1	25×256
dense_1	256
dense_2	256×6
dense_2	6

$time\_steps = 128$ and the other with $cell\_size = 10$ and the same $time\_steps$), a fully connected layer of 128 units and a ReLu activation, and a softmax output layer.

Results and Analysis

This subsection reports two sets of experimental studies. The first assesses the impact of those key hyper-parameters, $freq$, a, K, and m, on the two proposed strategies with the CNN for the 1@MNIST dataset. The other set compares four algorithms focusing on the accuracy and communication cost with the LSTM for the HAR task. Thus, the HAR is more challenging than MNIST.

Effect of the Parameters

Together with related discussions, the experiments aim to offer a basic understanding of hyper-parameter settings rather than a detailed sensitivity analysis. We hope that the insights gained from the experiments can benefit industrial practitioners. Table 2.1 lists the hyper-parameters studied in the experiments.

When a particular hyper-parameter is under investigation, all others are set to be their default value. For instance, with regard to $freq = \{3/15, 5/15, 7/15\}$, the study sets $a = e/2$, $K = 20$, and $m = 2$.

This study adopts two metrics to measure the performance, namely **the best accuracy** within 200 rounds (the central model) and **the required rounds** for the central model reaching 95.0% (accuracy). Note that the same computing rounds suggest the same communication cost. Tables 2.4, 2.5, and 2.6 list the average (AVG) and standard deviation (STDEV) values of the results obtained over ten independent runs in all experiments. The average value is listed before the standard deviation in parenthesis in the tables.

Based on the results in Tables 2.4, 2.5, and 2.6, the following observations can be made.

- **Analysis on *freq*:** According to Table 2.4, fewer communication costs will be required with a lower exchange frequency, which is very much expected. However, the accuracy deteriorates as $freq$ becomes too low.

Table 2.4 Experimental results on $freq$

Freq	Accuracy[a]	Round[b]
3/15	96.72% (0.0037)	115.74 (32.19)
5/15	97.08% (0.0037)	87.82 (20.03)
7/15	97.12% (0.0046)	95.43 (14.64)

[a] AVG (STDEV) of best overall accuracy within 200 rounds.
[b] ASTW_FedAVG on 1@MNIST reaches the accuracy 95.0% within 200 rounds

Table 2.5 Experimental results on a

a	Accuracy[a]	Round[b]
e[c]	96.92% (0.0041)	75.26 (2.09)
$e/2$	97.08% (0.0037)	87.82 (20.03)

[a] AVG (STDEV) of best overall accuracy within 200 rounds.
[b] ASTW_FedAVG on 1@MNIST reaches the accuracy 95.0% within 200 rounds.
[c] $e \approx 2.72$

- **Analysis on a**: Model aggregation adopts a to control the time effect. With a being set to e, the more recently updated local models are more heavily weighted in the aggregated model, and a takes the value $e/2$, the previously updated local models will have a greater impact on the central model. Table 2.5 shows that $e/2$ is a better option for the CNN on the 1@MNIST dataset. When $a = 1$, the algorithm turns into AS_FedAVG, meaning that the parameters uploaded in different rounds will be equally important in the aggregation.

K and m are two hyper-parameters reflecting the scalability of federated learning. The CNN-based MNIST tasks assess different combinations of the two parameters and calculate the AVG and STDEV values with the recognition accuracies. According to Table 2.6 summarizing those results, the following three conclusions can be made.

First, the more participating clients (the larger m) lead to the higher recognition accuracy. Second, as in the cases corresponding to the first, third, fourth, and fifth rows in the table, ASTW_FedAVG shows more competitive performance than FedAVG. Third, FedAVG comes with a slight lead when the total number of clients (K) is smaller, and C is higher, according to the second row of the table. This implies that

Table 2.6 Experimental results on the scalability of the algorithm

Scalability		FedAVG	ASTW_FedAVG
K	m		
10	1	94.26% (0.0083)	**94.76% (0.0102)**
10	2	**96.16% (0.0167)**	96.11% (0.0909)
20	2	96.83% (0.0097)	**97.16% (0.0096)**
30	3	96.50% (0.0020)	**97.79% (0.0067)**
30	9	97.14% (0.0029)	**98.22% (0.0070)**

* AVG (STDEV) of best overall accuracy within 200 rounds

the advantage of the proposed algorithm over the traditional federated learning will become more apparent as the number of clients increases. Most real-world problems typically involve a large number of clients, which makes the third encouraging.

Comparison of the accuracy and communication cost
Comparisons of the overall performance with hyper-parameters being set as default values are carried out to evaluate all algorithms under comparison. In the experiments, we generate five local datasets, respectively, for MNIST and HAR. For instance, 1@MNIST corresponds to the local dataset 1 of task MNIST. The importance of the temporal weighting is first demonstrated by comparing the changes of the accuracy over the computing rounds, i.e., that between the baseline FedAVG and TW_FedAVG without an asynchronous update. Figures 2.7 and 2.8, respectively, show the results of the MNIST and HAR experiments.

The following conclusions can thus be drawn. First, the proposed temporally weighted aggregation helps the central model to converge to an acceptable accuracy. For datasets 1@MNIST and 2@MNIST, TW_FedAVG requires about 30 communication rounds to reach an accuracy of 95.0%, while the traditional FedAVG needs about 75 rounds, leading to a reduction of 40%. Similar conclusions can also be drawn on datasets 3@MNIST, 4@MNIST, and 5@MNIST, although the accuracy of TW_FedAVG becomes more fluctuating. Second, for the more challenging HAR task, TW_FedAVG can achieve a higher accuracy than FedAVG except on 5@HAR. Notably, TW_FedAVG only requires about 75 rounds to achieve an accuracy of 90%, while FedAVG requires around 750, resulting in a nearly 90% reduction. Even on 5@HAR, TW_FedAVG shows a much faster convergence than FedAVG in the early stage. Finally, as shown by the training curve, the temporally weighted aggregation may cause some fluctuations, which may be attributed to the fact that the contributions of some high-quality local models are less weighted in some communication rounds.

Finally, Table 2.7 summarizes the comparisons of the four approaches on ten test cases (MNIST and HAR). The metrics under consideration are the required number of **rounds**, the classification **accuracy** (listed in parenthesis), and the total communication cost (**C. Cost** for short). With them, the following observations can be made:

- In most cases, either ASTW_FedAVG or TW_FedAVG outperforms FedAVG in terms of all metrics.
- TW_FedAVG performs the best (the total number of rounds and the best accuracy) in most cases. The temporally weighted aggregation strategy contributes to convergence acceleration and performance enhancement.
- Considering the total communication cost, ASTW_FedAVG holds a slight lead over TW_FedAVG on MNIST, whereas TW_FedAVG outperforms ASTW _FedAVG on HAR. The layer-wise asynchronous model update strategy dramatically reduces the single-round communication cost.
- AS_FedAVG performs the worst among the four compared algorithms. Compared with the performance of the AS_FedAVG only adopting the asynchronous strategy

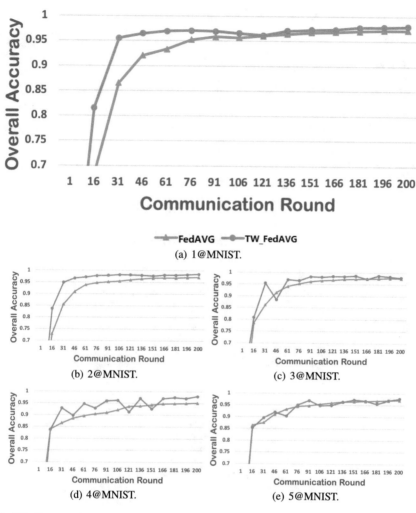

Fig. 2.7 Comparative studies on the temporally weighted aggregation on dataset MNIST using the CNN

and the ASTW_FedAVG using both of them, the asynchronous one always needs the help of the temporally weighted aggregation.

2.4   Trained Ternary Compression for Federated Learning

In this section, we will introduce a quantization based protocol for communication cost reduction in horizontal federated learning, which can be regarded as a direct application of model compression in federated learning.

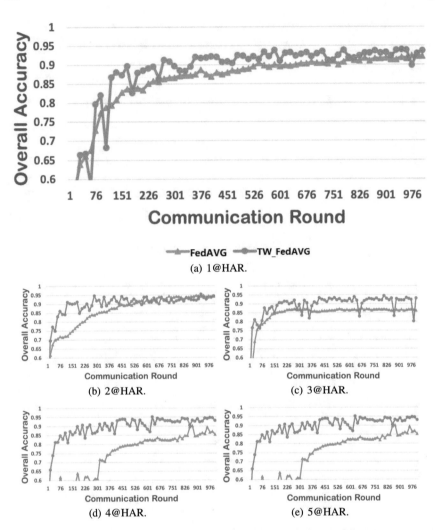

Fig. 2.8 Comparative studies on the temporally weighted aggregation on dataset HAR using the LSTM

2.4.1 Binary and Ternary Compression

Usually, variables in a neural network are stored as 32-bit or 64-bit floating-point numbers. For example, a 32-bit floating-point number can be represented by $(-1)^s \times m \times 2^{(e-127)}$, where s is the sign bit, m is the 23-bit mantissa and e is the 8-bit exponent, respectively, as shown in Fig .2.9.

Table 2.7 Experiments on performance

Dataset_ID @Task	FedAVG		TW_FedAVG		ASTW_FedAVG		AS_FedAVG	
	Rnd. (Acc.)*	C. Cost**	Rnd. (Acc.)*	C. Cost**	Rnd. (Acc.)*	C. Cost**	Rnd. (Acc.)*	C. Cost**
1@MNIST*	75 (97.2%)	6.16	**31** **(97.9%)**	0.74	106 (97.7%)	1	175 (95.4%)	2.46
2@MNIST*	85 (97.2%)	6.76	**32** **(98.5%)**	1.33	61 (98.1%)	1	–(94.8%)†	–†
3@MNIST*	73 (97.7%)	5.99	**31** **(98.7%)**	1.13	70 (97.9%)	1	–(94.9%)†	–†
4@MNIST*	196 (95.2%)	8.18	**61** **(98.0%)**	1.14	136 (96.1%)	1	–(93.1%)†	–†
5@MNIST*	98 (97.1%)	3.28	**76** **(97.8%)**	3.17	61 (96.1%)	1	109 (96.7%)	2.66
1@HAR**	526 (92.3%)	4.72	**166** **(94.0%)**	**0.69**	358 **(94.6%)**	1	–(89.2%) ‡	–‡
2@HAR**	451 (94.7%)	5.65	**119** **(95.4%)**	**0.57**	313 **(95.9%)**	1	–(82.9%) ‡	–‡
3@HAR**	–(87.3%) ‡	–‡	**174** **(94.4%)**	**0.69**	376 (93.2%)	1	–(88.5%) ‡	–‡
4@HAR**	856 (90.2%)	7.05	**181** **(95.1%)**	1.28	(94.1%)	1	751 (91.2%)	5.30
5@HAR**	571 (92.6%)	5.49	**155** **(93.6%)**	**0.57**	404 (93.1%)	1	646 (92.5%)	2.38

*Cells are filled when the accuracy reaches 95% within 200 rounds.
**Cells are filled when the accuracy reaches 90% within 1000 rounds.
†Cells are marked when it can not reach 95% within 200 rounds.
‡Cells are marked when it can not reach 90% within 1000 rounds.
*Rounds are needed to reach a certain accuracy, after which the best accuracy (within 200/1000 rounds) is listed in parenthesis.
**Total communication cost; the communication cost of ASTW_FedAVG is set to 1 for normalization

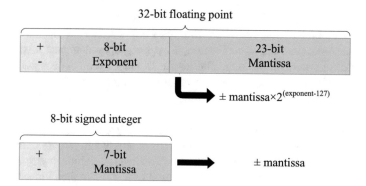

Fig. 2.9 The number representation of 32-bit floating point and 8-bit signed integer

Linear quantizer

Nonlinear quantizer

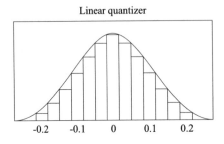
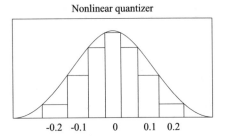

Fig. 2.10 Linear and nonlinear quantization

An 8-bit signed integer in the range −128 to 127 consumes only 1/4 the number of bits compared to 32-bit floating point. Intuitively, if mapping a 32-bit real number r in a vector to an 8-bit signed integer q with a scaling factor S and a zero point Z (usually set to 0), 3× space can be saved [29]. The mapping process is as follows:

$$r = S(q - Z), \tag{2.3}$$

$$S = \frac{r_{max} - r_{min}}{q_{max} - q_{min}}. \tag{2.4}$$

where r_{max} & r_{min}, q_{max} & q_{min} are the maximum and minimum values of r and q, respectively.

This strategy, often referred to as quantization or fake quantization, has been adopted in many popular deep learning frameworks, such as TensorFlow [29] and PyTorch [30]. In general, quantization is the process of reducing the precision of operations and operands to save energy and space costs, which aims to minimize the reconstruction error between the full-precision variable and the quantized one [2]. Quantization becomes a mainstream method for model compression for its effective and simpleness. The two most common quantization approaches, as shown in Fig. 2.10, are linear quantization (or uniform quantization) and nonlinear quantization. The distance between different levels remains constant after a linear quantization, making it simple and straightforward to implement. To alter the distance between different levels in nonlinear quantization, a mapping approach such as the natural logarithmic function can be used. The levels of quantization reflect the precision and number of bits required to map the variable during compression. For example, Dettmers adopts 8-bit quantization to compress weight and activation values in deep learning models [31].

The precision can be reduced even more aggressively to a 1 bit (i.e., −1 and +1) or 2 bits (i.e., −1, 0 and +1), which are often referred to binary and ternary compression. Courbariaux et al. [32] compress the weights, activation values, and gradients of the full-precision model into a matrix composed of 0 and 1, called binary network. However, such aggressive quantization leads to significant performance degradation.

To alleviate this problem, Li et al. [33] propose a ternary weight network (TWN), which quantizes the full-precision w as the product of the ternary matrix w^t composed of $-1, 0, +1$ and a scaling factor α. The objective of quantization is as follows,

$$\alpha^*, w^{t*} = \arg\min_{\alpha, w^t} \|w - \alpha w^t\|_2^2, \tag{2.5}$$

where α^* and w^{t*} are the optimal solutions that hold $w \approx \alpha^* w^{t*}$. And in practice, the quantization of matrix w^{t*} is a layer-wise operation:

$$w_l^t = \begin{cases} +1, & w_l > \Delta_l \\ 0, & |w_l| \leq \Delta_l \\ -1, & w_l < -\Delta_l, \end{cases} \tag{2.6}$$

in which w_l and w_l^t are the full-precision and quantized weight matrix of the lth layer respectively, and $w = \{w_1, w_2, \ldots, w_l, \ldots\}$, $w^t = \{w_1^t, w_2^t, \ldots, w_l^t, \ldots\}$. Δ_l is the layer-wise threshold, and the authors provide an empirical rule to compute its optimal value,

$$\Delta_l^* = \frac{0.7}{d^2} \sum_i^{d^2} (|w_l^i|), \tag{2.7}$$

where d is the dimension of w_l. Then the optimal solution of the scaling factor α can be derived from

$$\alpha_l^* = \frac{1}{|I_{\Delta_l}|} \sum_i^{|I_{\Delta_l}|} |w_i|, \tag{2.8}$$

where $I_{\Delta_l} = \{i \mid w_i > \Delta_l\}$.

2.4.2 Trained Ternary Compression

Empirically, the scaling factor α helps reduce the Euclidean distance between w and w^{t*} [33]. However, the empirical solution of α^* of TWN may lead to unstable performance in some cases. To overcome this problem and improve the performance of quantized deep networks, Zhu et al. [34] propose a trained ternary quantization algorithm (TTQ for short), using the gradient-based method to optimize two quantization factors (positive factor $\alpha_{p,l}$ and negative factor $\alpha_{n,l}$) to scale the three-valued matrix in each layer.

The overall workflow of TTQ is given in Fig. 2.11, where the normalized full-precision weights are quantized by Δ_l, $\alpha_{p,l}$ and $\alpha_{n,l}$:

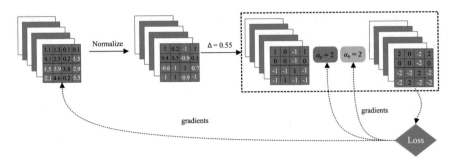

Fig. 2.11 An example of trained ternary quantization

$$w_l^t = \begin{cases} +1 \times \alpha_{p,l}, & w_l > \Delta_l \\ \quad\quad 0, & |w_l| \leq \Delta_l \\ -1 \times \alpha_{n,l}, & w_l < -\Delta_l, \end{cases} \tag{2.9}$$

and TTQ computes Δ_l using the following heuristic method,

$$\Delta_l = threshold \times \max(|w_l|), \tag{2.10}$$

where $\Delta = \{\Delta_1, \Delta_2, \ldots, \Delta_l, \ldots\}$, and *threshold* is the empirical constant usually set to 0.05.

2.4.3 Trained Ternary Compression for Federated Learning

In this section, we first propose a federated trained ternary quantization (FTTQ for short) to reduce the energy consumption (computational resources) required for each client during inference and the upstream and downstream communication costs. Subsequently, a ternary federated averaging protocol (T-FedAvg for short) is suggested.

2.4.3.1 Federated Trained Ternary Quantization

One challenge remains when threshold-based quantization is to be implemented in a federated learning system: Since the sparsity of the quantized local models may greatly differ from client to client, the global model will be biased towards the local model having the maximum weight range, if the same threshold is used for all clients. To address this issue, we start by normalizing the weights to $[-1, 1]$:

$$w^s = g(w), \tag{2.11}$$

where w^s is the normalized full-precision weight matrix, and g is the scaling function that normalizes the elements of a vector to the range $[-1, 1]$. However, the amplitude imbalance problem [35] may occur when normalizing the entire matrix w instead of layer by layer, which may lead to a significant performance degeneration. Hence, layer-wise normalization is used in this method.

For simplicity, we drop the subscript l of the quantization factors, thresholds and weights hereafter. Thus, we calculate the quantization threshold Δ according to the normalized weights as follows:

$$\Delta = T_k \times \max(|w^s|), \tag{2.12}$$

where T_k is a hyper-parameter for kth client, and rand $(0, 1)$ is the operation to generate a uniformly distributed random number in $[0, 1)$. Note that we introduce randomness into T_k to make it more difficult to reverse-engineer the weights. We limit the randomness to a controlled range to avoid a significant loss in performance:

$$T_k = \begin{cases} 0.05 + 0.01 \times \text{rand}(0, 1), & \text{if } \text{rand}(0, 1) > 0.5 \\ 0.05 + 0.01 \times \dfrac{k}{K}, & \text{if } \text{rand}(0, 1) \leq 0.5 \end{cases} \tag{2.13}$$

According to the Eqs. (2.11)–(2.12), it can be inferred that the values of Δ in some layers of the normalized network will be very close. Inspired by [33], we propose an alternate strategy to determine the threshold based on sparsity:

$$\Delta = \frac{T_k}{d^2} \sum_i^{d^2} (|w_i^s|), \tag{2.14}$$

where d is the dimension of the matrix w_i^s, and Δ is calculated layer-wise. This makes the weights difficult to invert, but it can also be viewed as an extension of Eq. (2.12) because

$$\Delta = T_k \times \frac{1}{d^2} \sum_i^{d^2} (|w_i^s|) \leq T_k \times \frac{1}{d^2} (d^2 \times \max |w^s|) \tag{2.15}$$
$$\leq T_k \times \frac{1}{d^2} (d^2 \times 1) \leq T_k.$$

and if we set the value of T_k to 0.7, then Δ is equal to the optimal value suggested in the literature [33].

Once the threshold is determined, a step function ε is adopted to map the element whose absolute value is greater than Δ to 1, and the rest to 0, that is, to get the *mask*:

$$mask = \varepsilon(|w^s| - \Delta), \tag{2.16}$$

and *mask* can be rewritten as the union of the positive index matrix I_p and negative index matrix I_n:

$$I_p = \{i \mid w_i^s > \Delta\}, \tag{2.17}$$

$$I_n = \{i \mid w_i^s < -\Delta\}. \tag{2.18}$$

Then, w^s can be converted into a ternary matrix I_t consisting of $-1, 0, +1$ through the sign function,

$$I_t = \text{sign}(mask \odot w^s), \tag{2.19}$$

where $\odot$ is the Hadamard product, or point-to-point product.

Different from the vanilla TTQ, a single layer-wise factor α_q is used for quantization instead of two positive and negative factors α_p and α_n:

$$w^t = \alpha_q \times I_t, \tag{2.20}$$

since we have demonstrated experimentally and theoretically that a single quantization factor can also achieve competitive performance. In addition, reducing one gradient-based factor improves the computational efficiency of large-scale federated learning systems.

However, the derivative of the quantized model w^t with respect to w, that is, $\frac{\partial w^t}{\partial w}$, may become infinity or zero, making the entire network hard to train using the gradient based method, since

$$\frac{\partial \mathcal{L}}{\partial w} = \frac{\partial \mathcal{L}}{\partial w^t} \cdot \frac{\partial w^t}{\partial w}. \tag{2.21}$$

To avoid this problem, Zhu et al. [34] propose the following method to calculate the gradients of α_q and w,

$$\frac{\partial \mathcal{L}}{\partial \alpha_q} = \sum_{i \in I_p} \frac{\partial \mathcal{L}}{\partial w_i^t}, \tag{2.22}$$

$$\frac{\partial \mathcal{L}}{\partial w} = \begin{cases} 1 \times \dfrac{\partial \mathcal{L}}{\partial w^t}, & |w| \leq \Delta, \\[2mm] \alpha_q \times \dfrac{\partial \mathcal{L}}{\partial w^t}, & \text{otherwise.} \end{cases} \tag{2.23}$$

The pseudo code of FTTQ is summarized in Algorithm 5. It can be seen that FTTQ compresses the client model from 32 bits to about 2 bits, significantly reducing the cost of upstream communication.

However, the weights of the global model become real-valued again after aggregation, which means that the downstream communication cost does not decrease if no additional strategies are employed. To address this problem, a server-side ternary federated averaging strategy will be presented in Sect. 2.4.3.2.

Algorithm 5 Federated Ternary Quantization (FTTQ)

Input: Full-precision model w, quantization factor α_q, loss function l, dataset $\mathcal{D}$ composed of sample pairs $(\boldsymbol{x}_i, y_i)$, $i = \{1, 2, ..., |\mathcal{D}|\}$, learning rate η
Output: Quantized model w^t
1: **for** $(\boldsymbol{x}_i, y_i) \in \mathcal{D}$ **do**
2:   $w^s \leftarrow g(w)$
3:   $mask \leftarrow \varepsilon(|w^s| - \Delta)$
4:   $I_t = \text{sign}(mask \odot w^s)$
5:   $w^t \leftarrow \alpha_q \times I_t$
6:   $\mathcal{L} \leftarrow l(\boldsymbol{x}_i, y_i; w^t)$
7:   $\alpha_q \leftarrow \alpha_q + \eta \frac{\partial \mathcal{L}}{\partial \alpha_q}$
8:   $w \leftarrow w + \eta \frac{\partial \mathcal{L}}{\partial w}$
9: **end for**
10: **Return** w^t (including α_q, I_t)

2.4.3.2 Ternary Federated Averaging

The overall workflow of the ternary federated averaging (T-FedAvg) is shown in Fig. 2.12. First, the client normalizes and quantizes the local model using FTTQ. Then, each participated client uploads the quantized model and quantization factor vector to the server, and the server aggregates all local models to obtain a full-precision global model. Finally, the server quantizes the global model again and broadcasts the quantized global model back to all clients.

2.4.3.3 Upstream Communication

Let $\mathbb{K} = \{1, 2, ..., |\lambda K|\}$ be the index set of participating clients, where λ is the participation ratio and K is the total number of clients in a federated learning system. The full-precision and quantized models of client $k \in \mathbb{K}$ are represented by w_k^s and w_k^t, respectively. In one communication round, client k uses the local dataset $\mathcal{D}_k$ for training and quantization, and then uploads the quantized weight w_k^t (composed of α_q and I_t) to the server. In the inference phase, only the quantized model is used for prediction.

In this way, the upstream communication cost can be reduced significantly. For example, if we configure a federated learning system involving 40 clients and the global model that requires 25 Mb of storage, the upstream communication cost for the standard federated learning is about 1 Gb per round. By contrast, our method reduces the cost per round to 65 Mb, about 1/16 of the standard method.

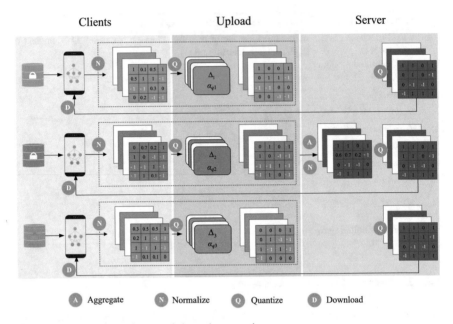

Fig. 2.12 The flowchart of ternary federated aggregation

2.4.3.4 Downstream Communication

Similarly, in each round the server restores the uploaded quantized local models to full-precision models, and aggregates them to obtain the global model. Then, we design two different strategies to broadcast the global model back to all clients.

Strategy I: The server uses the threshold $\Delta^S = 0.05 \times \max(|w_r|)$ and two server quantization factors

$$\alpha_p^S = \frac{1}{|I_p^S|} \sum_i^{|I_p^S|} (|w_r^i|), \tag{2.24}$$

$$\alpha_n^S = \frac{1}{|I_n^S|} \sum_i^{|I_n^S|} (|w_r^i|), \tag{2.25}$$

to quantize the global model, where r denotes the index of the communication round, I_p^S and I_n^S are determined by the Eqs. (2.17)–(2.18). In the end, the server broadcasts the quantized global model to the clients.

Strategy II: If the performance of the quantized model drops by more than 3%, the quantized model is considered to have collapsed and cannot be used, and the server will send the aggregated full-precision model w_{r+1} to all clients.

By quantizing the global model, communication cost in the download phases can be greatly saved, which also brings advantages when deploying deep neural networks on resource-constrained devices. Algorithm 6 summarizes the overall workflow of the proposed ternary federated aggregation. And the difference between strategy I and strategy II lies in the downstream communication cost.

Algorithm 6 Ternary Federated Aggregation (T-FedAvg)

Input: Initialize global model w_0
Init: Broadcast w_0 to all clients
1: **for** round r = 1,..., T **do**
2:     **for Client** k $\in \mathbb{K} = \{1, 2, ..., |\lambda K|\}$ **in parallel do**
3:         Load $\mathcal{D}_k$
4:         $w_k \leftarrow w_{r-1}^t$ or w_{r-1}
5:         Initialize α_q
6:         $w_{k,r}^t \leftarrow \textbf{FTTQ}(w_k, \alpha_q)$
7:         Upload $w_{k,r}^t$ to the server
8:     **end for**
9:     **procedure SERVER**(does)
10:         $w_r \leftarrow \sum_{k=1}^{\lambda K} \frac{|\mathcal{D}_k|}{\sum_{k=1}^{\lambda K} |\mathcal{D}_k|} w_{k,r}^t$
11:         $\Delta^S = 0.05 \times \max(|w_r|)$
12:         $I_p^S, I_n^S = \{i \mid w_{r,i} > \Delta^S\}, \{j \mid w_{r,j} < -\Delta^S\}$
13:         $\alpha_p^S, \alpha_n^S = \frac{1}{|I_p^S|} \sum_{i=1}^{|I_p^S|} (|w_{r,i}|), \frac{1}{|I_n^S|} \sum_{j=1}^{|I_n^S|} (|w_{r,j}|)$
14:         $w_r^t \leftarrow \alpha_p^S \times I_p^S - \alpha_n^S \times I_n^S$
15:         **if** w_r^t does not crash **then**
16:             Broadcast w_r^t $/*\text{Strategy I}*/$
17:         **else**
18:             Broadcast w_r $/*\text{Strategy II}*/$
19:         **end if**
20:     **end procedure**
21: **end for**

2.4.4 Theoretical Analysis

In this section, we theoretically analyze the convergence of two quantization factors α_p, α_n in TTQ, and then discuss the unbiasedness of FTTQ and the convergence speed of T-FedAvg. Finally, we present a theoretical proof that T-FedAvg can alleviate the weight divergence in federated aggregation.

2.4.4.1 Convergence of the Quantization Factors

To simplify the convergence proof of α_p, α_n, it is assumed that the following assumption is satisfied when the network is initialized:

Assumption 2.1 The elements in the normalized full-precision w are uniformly distributed between -1 and 1,

$$\forall w_i \in w, w_i \sim U(-1, 1). \tag{2.26}$$

Then we have the following proposition.

Proposition 2.1 *Given a one-layer online gradient system, each element of its parameters is initialized with a symmetric probability distribution centered at 0, e.g., $w_i \sim U(-1, 1)$, which is quantized by TTQ with two iteratively adapted factors α_p, α_n and a fixed threshold Δ, then we have:*

$$\lim_{e \to +\infty} \alpha_p = \lim_{e \to +\infty} \alpha_n, \tag{2.27}$$

Proof The converged α_p^* and α_n^* can be regarded as the optimal solution of the quantization factors, which can reduce the Euclidean distance between the full-precision weight w and the quantization weight w^t (equal to $\alpha_p I_p - \alpha_n I_n$), which is

$$\alpha_p^*, \alpha_n^* = \arg\min_{\alpha_p, \alpha_n} \; \|w - \alpha_p I_p + \alpha_n I_n\|_2^2, \tag{2.28}$$

where $I_p = \{i|\theta_i \geq \Delta\}$, $I_n = \{j|\theta_j \leq -\Delta\}$, $I_z = \{k||\theta_k| < \Delta\}$, and according to Eq. (2.6) we have

$$w - \alpha_p I_p + \alpha_n I_n = \begin{cases} w_i - \alpha_p, & i \in I_p \\ w_k, & k \in I_z \\ w_j + \alpha_n, & j \in I_n, \end{cases} \tag{2.29}$$

then the original problem can be transformed into

$$\begin{aligned} \|w - \alpha_p I_p &+ \alpha_n I_n\|_2^2 \\ &= \sum_{i \in I_p}(w_i - \alpha_p)^2 + \sum_{j \in I_n}(w_j + \alpha_n)^2 + \sum_{k \in I_z} w_k^2 \\ &= |I_p|\alpha_p^2 + |I_n|\alpha_n^2 - 2\alpha_p \sum_{i \in I_p} w_i + 2\alpha_n \sum_{j \in I_n} w_j + C, \end{aligned} \tag{2.30}$$

where $C = \sum_{i \in I_p} w_i^2 + \sum_{j \in I_n} w_j^2 + \sum_{k \in I_z} w_k^2$ is a constant independent of α_p and α_n, then the optimal solution to Eq. (2.30) can be obtained by

$$\alpha_p^* = \frac{1}{|I_p|} \sum_{i \in I_p} w_i,$$

$$\alpha_n^* = -\frac{1}{|I_n|} \sum_{j \in I_n} w_j,$$

(2.31)

and since the weights are distributed symmetrically, α_p^* and α_n^* will converge to the same value.

Then according to Eq. (2.31), α_p^* is determined by the elements in $I_p = \{i | w_i \geq \Delta\}$, where Δ is a fixed number once the parameters are generated under Assumption 2.1, hence the elements indexed by I_p obey a new uniform distribution between Δ and 1, that is

$$\forall i \in I_p, \ w_i \backsim U(\Delta, 1).$$

(2.32)

Therefore, the probability density function f of w_i ($i \in I_p$) can be expressed as $f(x) = \frac{1}{1-\Delta}$. Then, according to Proposition 2.1 and Eq. (2.31), we have

$$
\begin{aligned}
\mathbb{E}\left(\alpha_p^*\right) &= \mathbb{E}\left(\frac{1}{|I_p|} \sum_{i \in I_p} w_i\right) = \frac{1}{|I_p|} \mathbb{E}\left(\sum_{i \in I_p} w_i\right) \\
&= \frac{1}{|I_p|} |I_p| \int_\Delta^1 u f(u) du = \int_\Delta^1 u f(u) du \\
&= \frac{1 + \Delta}{2},
\end{aligned}
$$

(2.33)

where u is a random number obeying the Eq. (2.32), and $|I_p|$ represents the number of elements in I_p.

This completes the proof. □

2.4.4.2 Unbiasedness of FTTQ

Under Sect. 2.4.4.1 and Assumption 2.1, we can prove the following proposition:

Proposition 2.2 *Let w be the local scaled network parameters defined in Assumption 2.1 of one client in a given federated learning system. If w is quantized by the FTTQ algorithm, then we have*

$$\mathbb{E}[FTTQ(w)] = \mathbb{E}(w).$$

(2.34)

Proof We see that

$$
\begin{aligned}
\mathbb{E}[FTTQ(w)] &= \mathbb{E}\left[\alpha_q^* \times \text{sign}(mask(w) \odot w)\right] \\
&= \mathbb{E}(\alpha_q^*)\mathbb{E}\left[mask(w) \odot \text{sign}(w)\right],
\end{aligned}
$$

(2.35)

and

$$\mathbb{E}\left[mask(w) \odot sign(w)\right] = P\left[mask(w) = 1\right] \times 1 + P\left[mask(w) = 0\right]$$
$$\times 0 + P\left[mask(w) = 1\right] \times (-1)$$
$$= \frac{1 - \Delta}{2} \times 1 + \Delta \times 0 + \frac{1 - \Delta}{2} \times (-1) \quad (2.36)$$
$$= 0,$$

and since

$$\mathbb{E}\left(\alpha_q^*\right) = \mathbb{E}\left(\alpha_p^*\right) = \frac{1 + \Delta}{2}, \quad (2.37)$$

then

$$\mathbb{E}\left[FTTQ(w)\right] = \mathbb{E}(\alpha_q^*)\mathbb{E}\left[mask(w) \odot sign(w)\right]$$
$$= \frac{1 + \Delta}{2} \times 0 = 0, \quad (2.38)$$

and with Assumption 2.1, we have

$$\mathbb{E}\left(w\right) = \frac{1 + (-1)}{2} = 0, \quad (2.39)$$

therefore

$$\mathbb{E}\left[FTTQ(w)\right] = \mathbb{E}\left(w\right). \quad (2.40)$$

Proof completes.   □

Hence, the FTTQ quantizer output can be considered as an unbiased estimator of the input [36]. We can guarantee the unbiasedness of FTTQ in federated learning systems when the weights are uniformly distributed.

2.4.4.3 Convergence Rate of T-FedAvg

If the unbiasedness of FTTQ holds, we can have the following proposition to analyze the convergence rate of T-FedAvg.

Proposition 2.3 *Let $\mathcal{L}_1, ..., \mathcal{L}_K$ be L-smooth and μ-strongly convex objective functions, $\mathbb{E}||\nabla \mathcal{L}_k(w_{k,r}, \xi_{k,r}) - \nabla \mathcal{L}_k(w_{k,r})||^2 \leq \delta_k^2$, $\mathbb{E}||\nabla J_k(w_{k,r})||^2 \leq G^2$, where $\xi_{k,r}$ is a mini-batch sampled uniformly at random from kth client's data. Then, for a federated learning system with K devices (full participation) and IID data distribution, the convergence rate of T-FedAvg is $O(\frac{1}{KR})$, where R is the total number of SGD iterations performed by each client.*

Proof Here, we give a short proof based on the literature [37–39].
    Let $\hat{w}_R = \sum_{k=1}^{K} \frac{|\mathcal{D}_k|}{\sum_{k=1}^{K} |\mathcal{D}_k|} w_k$, $\hat{w}_R^t = \sum_{k=1}^{K} \frac{|\mathcal{D}_k|}{\sum_{k=1}^{K} |\mathcal{D}_k|} FTTQ(w_k)$. Under Proposition 2.2, we have

$$E(\hat{w}_R^t) = \sum_{k=1}^{K} \frac{|\mathcal{D}_k|}{\sum_{k=1}^{K} |\mathcal{D}_k|} \mathbb{E}[FTTQ(w_k)]$$

$$= \sum_{k=1}^{K} \frac{|\mathcal{D}_k|}{\sum_{k=1}^{K} |\mathcal{D}_k|} \mathbb{E}(w_k) \qquad (2.41)$$

$$= \mathbb{E}(\hat{w}_R).$$

If we sett $v = \max_k K \frac{|\mathcal{D}_k|}{\sum_{k=1}^{K} |\mathcal{D}_k|}$, $\kappa = \frac{L}{\mu}$, $\gamma = max\{32\kappa, E\}$ and step size $\eta_r = \frac{1}{4\mu(\gamma+r)}$, then according to Proposition 2.3, we have

$$\mathbb{E}(\mathcal{L}(\hat{w}_r^t)) - \mathcal{L}^* = \mathbb{E}(\mathcal{L}(\hat{w}_r) - \mathcal{L}(w^*)) \leq \frac{L}{2} \mathbb{E}||\hat{w}_r - w^*||^2. \qquad (2.42)$$

Then, according to Eqs. (2.41)–(2.42) and the proof given by Qu et al. [39], the convergence rate of T-FedAvg is

$$\mathbb{E}(\mathcal{L}(\hat{w}_r^t)) - \mathcal{L}^* \leq \frac{L}{2} \mathbb{E}||\hat{w}_r - w^*||^2$$

$$\leq 2\hat{c}||w_0 - w^*||^2 \left[\frac{\kappa}{\mu} \frac{1}{K} \frac{1}{R+\gamma} v^2 \delta^2 + 96 \frac{\kappa^2}{\mu} \frac{1}{(R+\gamma)^2} E^2 G^2 \right] \qquad (2.43)$$

$$= O\left(\frac{\kappa}{\mu} \frac{1}{K} \frac{1}{R} v^2 \delta^2 + \frac{\kappa^2}{\mu} \frac{1}{R^2} E^2 G^2 \right),$$

where $\hat{c}$ is a constant large enough to be used for inequality scaling. Therefore, if we set the number of local iterations $E = O(\sqrt{R/K})$ then $O(E^2/R^2) = O(\frac{1}{KR})$, the convergence rates of FedAvg and T-FedAvg will be $O(\frac{1}{KR})$. Readers are referred to [37–39] for a detailed proof.

2.4.4.4 Weight Divergence

This section mainly discusses the weight divergence property of the proposed algorithm. The property was first proposed in [40] and widely used by researchers.

First, assume that the following initialization assumption is satisfied in each client of a federated learning system.

Assumption 2.2 When a federated learning system with K clients and one server is established, all clients will be initialized with the same global model. □

Zhao et al. [40] firstly propose a criterion to define the weight divergence of FedAvg, which is

$$WD_{Fed} = ||w^{Fed} - w^{Cen}||/||w^{Cen}||, \qquad (2.44)$$

where w^{Cen} is the model trained by centralized learning, and w^{Fed} is the full-precision model obtained by FedAvg. Then we have the following proposition.

Proposition 2.4 *Given one layer weight matrix with d dimension in the global model $w^{Fed} \backsim U(-1, 1)$ of a federated learning system, and a quantized model w^{TFed} using Algorithm 6, then the expected weight divergence of the quantized model will be reduced by $\frac{2\sqrt{2}-\sqrt{7}}{2\sqrt{3}||w^{Cen}||}d$, if d^2 is large enough.*

Proof According to Sect. 2.4.4.2, we can compute the weight divergence for T-FedAvg:

$$WD_{TFed} = ||w^{TFed} - w^{Cen}||/||w^{Cen}||, \tag{2.45}$$

then, the numerator of the Eq. (2.45) can be rewritten as

$$||w^{TFed} - w^{Cen}|| = ||\alpha_p^{S*} I_p^S - \alpha_n^{S*} I_n^S - w^{Cen}||, \tag{2.46}$$

and in Algorithm 6, if a fixed threshold is used and the elements are normalized to the interval $[-1, 1]$, $\Delta \approx 0.05$, which is very close to 0. Then under Assumption 2.1, we can calculate the limits of α_p^{S*} and α_n^{S*} when the number of elements in w^{Cen} is large enough, which is

$$\lim_{d^2 \to +\infty} \alpha_p^{S*} = \lim_{d^2 \to +\infty} \alpha_n^{S*} \approx \frac{1}{2}, \tag{2.47}$$

where d is the dimension of the matrix w^{Cen} (also known as the kernel size), since the size of w^{Cen} is the same as w^{TFed}, the index matrix I_p^S and I_n^S can also be used for w^{Cen}. Then if Eq. (2.47) holds, the expectation of the square of the weight divergence for T-FedAvg is equal to

$$
\begin{aligned}
\mathbb{E}\left(||w^{TFed} - w^{Cen}||_2^2\right) &= \mathbb{E}\left(||\frac{1}{2}I_p^S - \frac{1}{2}I_n^S - w^{Cen}||_2^2\right) \\
&= \mathbb{E}\left[\sum_{i \in I_p^S}\left(\frac{1}{2} - w_i^{Cen}\right)^2\right] + \mathbb{E}\left[\sum_{j \in I_n^S}\left(\frac{1}{2} + w_j^{Cen}\right)^2\right] \\
&= \frac{d^2}{2}\mathbb{E}\left[(\frac{1}{2} - w_i^{Cen})^2\right] + \frac{d^2}{2}\mathbb{E}\left[\left(\frac{1}{2} + w_j^{Cen}\right)^2\right],
\end{aligned} \tag{2.48}
$$

since w_i^{Cen}, $w_j^{Cen} \backsim U(-1, 1)$, we can define $u_1 \backsim U(-\frac{1}{2}, \frac{3}{2})$, $u_2 \backsim U(-\frac{3}{2}, \frac{1}{2})$ to represent $\frac{1}{2} - w_i^{Cen}$ and $\frac{1}{2} + w_j^{Cen}$, and the density of $f(u_1)$, $f(u_2)$ is equal to $\frac{1}{2}$. Therefore, Eq. (2.48) can be rewritten as

$$\frac{d^2}{2}\mathbb{E}\left[\left(\frac{1}{2}-w_i^{Cen}\right)^2\right]+\frac{d^2}{2}\mathbb{E}\left[\left(\frac{1}{2}+w_j^{Cen}\right)^2\right]$$

$$=\frac{d^2}{2}\int_{-\frac{1}{2}}^{\frac{3}{2}}u_1^2 f(u_1)du_1+\frac{d^2}{2}\int_{-\frac{3}{2}}^{\frac{1}{2}}u_2^2 f(u_2)du_2 \qquad (2.49)$$

$$=\frac{d^2}{2}\int_{-\frac{1}{2}}^{\frac{3}{2}}\frac{1}{2}u_1^2 du_1+\frac{d^2}{2}\int_{-\frac{3}{2}}^{\frac{1}{2}}\frac{1}{2}u_2^2 du_2=\frac{7}{12}d^2.$$

Similarly, we can derive the expectation of the numerator of the squared weight divergence of FedAvg by:

$$\mathbb{E}\left(||w^{Fed}-w^{Cen}||_2^2\right)=\mathbb{E}\left[\sum_{i\in I_p\cup I_n}(w_i^{Fed}-w_i^{Cen})^2\right], \qquad (2.50)$$

Consequently, we know that if two independent uniform distributions are on $[-1,\ 1]$, then their sum u_3 has the so-called triangular distribution on $[-2,\ 2]$ with density $f(u_3)=\frac{1}{4}(u_3+2)$ when $-2\le u_3\le 0$ and $f(u_3)=\frac{1}{4}(-u_3+2)$ when $0\le u_3\le 2$. Therefore, we have

$$\mathbb{E}\left[\sum_{i\in I_p^S\cup I_n^S}(w_i^{Fed}-w_i^{Cen})^2\right]$$

$$=\frac{d^2}{2}\int_{-2}^{0}\frac{1}{4}(u_3+2)u_3^2 du_3+\frac{d^2}{2}\int_{0}^{2}\frac{1}{4}(-u_3+2)u_3^2 du_3 \qquad (2.51)$$

$$=\frac{2}{3}d^2.$$

Proof completes.

Hence, T-FedAvg can alleviate the weight divergence of the federated learning system, and may alleviate the performance loss of federated learning under non-IID data.

2.4.5 Empirical Studies

In this section, we set up multiple controlled experiments to examine the performance of the proposed method against the standard federated learning algorithms in terms of test accuracy and communication cost.

2.4.5.1 Experimental Settings

To evaluate the performance of the proposed client quantization and ternary aggregation method in the federated learning system, we conduct experiments using both physical and simulated systems. The physical system consists of 1 CPU client and 5 GPU clients, which are wirelessly connected via local area network. The CPU client acts as a server to aggregate local models, and the remaining GPU clients are used for model training. The simulation system is set up with 100 clients. In all experiments, the clients only communicate with the server, and there is no information exchange between the clients.

(1) **Compared Algorithms**

- Baseline: A centralized algorithm that stores data in a data center and uses algorithms such as stochastic gradient descent (SGD) and Adam for parameter training.
- FedAvg: Classic and standard federated learning algorithm. The server performs the weighted aggregation of the local models. Each client is trained with its stored locally [41].
- CMFL: A communication-efficient federated learning framework. After local training, the difference between the old and new models helps each client determine whether or not to upload the model update [42].
- T-FedAvg: The proposed communication-efficient federated learning framework. Note that the first and last layers of the local model maintain full-precision.

(2) **Models**
MLP, CNN and ResNet*[1] networks are selected as shared models, representing three commonly used neural networks, namely fully connected, convolutional and residual convolutional, respectively. The detailed configuration is as follows:

- MLP: The model has one input layer, two hidden layers, and one output layer, with 784, 30, 20, and 10 neurons, respectively. There is no bias term, and ReLU is the activation function. The pretrained weights of the input layer have not been transmitted to the server to reduce the amount of communication.
- CNN: A shallow tiled convolutional neural network consisting of five convolutional layers and three fully connected layers. The output of the first

---

[1] The ResNet18 model is adopted, but the channels are adjusted from 64-128-256-512 to 64-64-64-64.

convolutional layer is processed by a ReLU function, the rest are followed by a batch normalization layer, a ReLU function and a max pooling layer.

- ResNet*: A simplified version of the widely used ResNet18 [43], where the number of input and output channels for all residual blocks is reduced to 64.

(3) **Datasets**

For the trials, two image classification benchmark datasets are employed, which are detailed in detail below:

- MNIST [44]: A 10-class grayscale handwriting dataset with 60,000 training samples and 10,000 test samples, each image has a dimension of 28×28. Since MNIST is easy to extract features, this dataset is mainly used to train small networks without data augmentation.
- CIFAR10 [45]: A widely used and hard-to-extract benchmark dataset consisting of 60,000 color images of 10 objects, 50,000 training samples and 10,000 testing samples. As suggested by the Wang et al. [46], using a mean and standard deviation of $\mu_r = 0.4914$, $\delta_r = 0.247$, $\mu_g = 0.4824$, $\delta_g = 0.244$, and $\mu_b = 0.4467$, $\delta_b = 0.262$, respectively, to normalize the three channels of RGB. Additionally, random cropping and horizontal random flipping are performed for data augmentation.

(4) **Data split**

The distribution of client data will greatly affect the overall performance of the federated learning system. In order to study the impact of data distribution, we generate data shards with different distributions:

- IID data: Each client stores an IID subset of the entire dataset, and all clients have the same number of samples.
- Non-IID data: Label Non-IID, that is, the union of samples in all clients is the entire dataset, but the number of classes contained in each client is not equal to the total number of classes in the entire dataset. Let $N_c (\leq 10)$ be the number of classes to which a client contains samples. In the extreme Non-IID case, N_c is equal to 1, which is usually not considered because if only one class is stored on each local device, no training (e.g. classification) is required.
- Unbalanced data: In general, dataset sizes on different clients vary widely. In order to study the impact of dataset size on federated learning systems, we control the number of samples per client in the experiments.

(5) **Hyper-parameter Configuration**

The basic configuration of the federated learning system is defined as follows:

- Number of clients: $K = 100$ in the MNIST simulation environment. $K = 5$ in the CIFAR10 physical environment.
- Participation ratio: For MNIST training, $\lambda = 0.1$. For CIFAR10 dataset, $\lambda = 1$.
- Class number in IID cases: $N_c = 10$.
- Number of local epochs: For MNIST, $E = 5$. For CIFAR10, $E = 10$.

Table 2.8 Models and hyperparameters

Models	MLP	CNN	ResNet*
Dataset	MNIST	CIFAR10	CIFAR10
Optimizer	SGD	Adam	Adam
Learning rate	0.01	0.001	0.008
Basic accuracy (%)	92.25 ± 0.06	85.63 ± 0.10	88.80 ± 0.20

The centralized algorithm and the federated learning algorithm use the same learning rate for CIFAR-10, and the learning rate is decayed every five rounds with a rate of 0.95. The number of samples per client is equal to the size of the dataset divided by K, and the batch size is fixed at 64.

The performance of each algorithm is assessed after 100 training communication rounds with the same number of training samples for a fair comparison, and the experimental results are the average of five independent runs. Taking MNIST as an example, there are 100 clients, each client has 600 samples, and the number of communication round is set to 100. Correspondingly, 60,000 samples are used to train the centralized method for 100 epochs. It should be pointed out that since the participation rate λ is less than or equal to 1, the actual number of training samples used in federated learning is always smaller than centralized learning. The hyperparameters used by the baseline are summarized in Table 2.8.

2.4.5.2 Results on IID Data

In this section, we examine the communication costs and performance of MLP, CNN, and ResNet* of three federated learning and one centralized learning approaches on IID MNIST and CIFAR10. Specifically, for the MNIST dataset, with a total of 100 clients, each client holds 600 samples of IID distribution, and λ is set to 0.1. For CIFAR10, each client stores 10,000 IID training samples, 5 clients in total, $\lambda = 1$.

First, we compare the communication costs of FedAvg, CMFL, and T-FedAvg for a fixed number of rounds. For T-FedAvg, the upstream communication cost is estimated by treating the three-valued part of the local model as 2-bit, and the downstream communication cost is computed by treating the three-valued part of the global model as 2-bit.

$$\text{download} = \left(1 - \frac{S_I}{T}\right) \times \text{upload}_f + \frac{S_I}{T} \times \text{upload}_t, \qquad (2.52)$$

where T is the total number of communication rounds in Algorithm 6, and S_I is the number of times the server executes strategy I. The results are listed in Table 2.9, where upload_f and upload_t are the upstream communication costs for FedAvg and T-FedAvg, respectively.

Table 2.9 The communication costs in 100 rounds on IID data. For MLP, 10 out of 100 clients (λ = 0.1) participate in each round of training. For CIFAR10, all five clients participate in each round of training ($\lambda = 1.0$)

Methods	MLP	CNN	ResNet*
	Upload/download (Mb)	Upload/download (Mb)	Upload/download (Mb)
FedAvg	19.53/19.53	16196.59/16196.59	9253.81/9253.81
CMFL	5.63/19.53	4079.75/16196.59	2666.80/9253.81
T-FedAvg	2.36/2.36	2201.84/3041.53	1229.27/6847.82

Table 2.10 Test accuracies achieved and weights width of different algorithms when trained on IID data

Accuracy (%)	MNIST	CIFAR10		Width
	MLP	CNN	ResNet*	Bit
Baseline	92.25 ± 0.06	85.63 ± 0.10	88.80 ± 0.20	32
FedAvg	90.63 ± 0.17	85.47 ± 0.14	88.34 ± 0.40	32
CMFL	90.91 ± 0.36	84.23 ± 0.71	87.13 ± 0.66	32
T-FedAvg	91.95 ± 0.11	84.46 ± 0.17	87.87 ± 0.18	2

It can be seen that the server does not use strategy II when training the MLP, so the communication cost of T-FedAvg in the upload and download phases is reduced by 88% compared to the standard FedAvg. When training CNN, the upload communication cost of T-FedAvg is reduced to 13%, and the download communication cost is 82% of FedAvg and CMFL, which means that the server executes strategy II multiple times to maintain the performance. Similarly, on the CIFAR10 dataset, T-FedAvg compresses nearly 94% upstream communication, 25% downstream communication, and outperforms CMFL in training ResNet*. It is worth noting that we did not transfer the parameters of the first layer in the MLP to the server.

Table 2.10 summarizes the performance of the four algorithms on the test sets. On the MNIST dataset, the accuracy of the MLP model based on centralized training reaches 92.25%. By contrast, models trained with CMFL and T-FedAvg achieve 90.91 and 91.95% accuracy, respectively, both outperforming FedAvg but slightly worse than the centralized approach. This is because the federated learning algorithms are trained with less valid data than the centralized algorithms when the participation ratio is less than 1. Interestingly, T-FedAvg outperforms FedAvg, probably because the ternary operation acts as a regularizer when training the MLP.

As for CIFAR10, the accuracy of the CNN model trained on T-FedAvg reaches 84.46%, compared to 85.47% of FedAvg. Correspondingly, when using ResNet*, the test accuracies are 87.87% and 88.34%, respectively. The test accuracy of CMFL is slightly worse than that of the other two algorithms, which may be affected by the reduction of communication cost. Due to the influence of quantization error, T-FedAvg performs slightly worse than FedAvg with a limited number of weights.

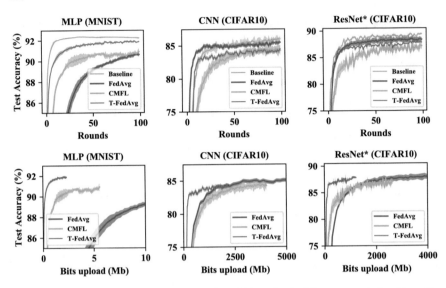

Fig. 2.13 The convergence curves over rounds and bits upload of the compared algorithms. For each algorithm, the solid line denotes the averaged test accuracy over five independent runs, and the shaded region denotes the standard deviation

However, the model size of the ternary part in the global and local models of T-FedAvg is only 1/16 of the full-precision model, and CMFL also reduces a large number of communication rounds. Therefore, the performance of these two methods can be considered to be satisfactory.

To compare the convergence performance of the four algorithms, Fig. 2.13 plots the convergence curves of each method. We can find that T-FedAvg converges faster than CMFL in all test instances, although FedAvg converges slightly faster than T-FedAvg when training a CNN on CIFAR10. It should be noted that since T-FedAvg quantizes the network during the training, it requires the least amount of communication to achieve the same test accuracy compared to CMFL and FedAvg. Furthermore, all federated learning methods converge more slowly than the centralized learning algorithms.

2.4.5.3 Results on Non-IID Data

Since the communication cost on the Non-IID data is basically the same as on IID data (refer to Table 2.9), here we focus on the impact of Non-IID data on the classification performance.

In federated learning, when N_c is less than the total number of classes in the training data, the data distribution can be considered as label Non-IID. The data distribution used in the following experiments is shown in Fig. 2.14, where the y-axis represents the sample labels (0–9). As shown in plot right, the original distribution

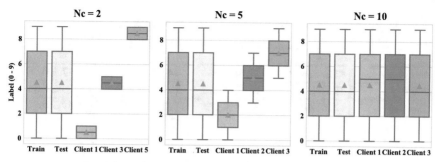

Fig. 2.14 Data distributions with different N_c settings. When $N_c = 2$, there is no overlap in data between clients and each client contains two categories (left). When $N_c = 5$, the samples on 10 clients are sampled by labels but there are some overlap between clients (middle). When $N_c = 10$, the samples on 10 clients are generated by randomly sampling (right)

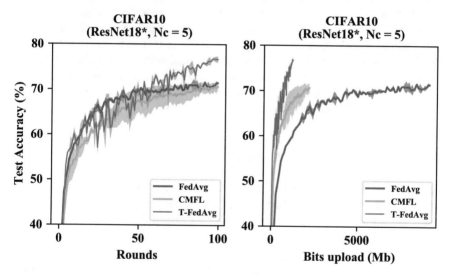

Fig. 2.15 Convergence trends and communication costs of FedAvg, CMFL and T-FedAvg when $N_c = 5$

of training and testing data is IID. That is, when N_c equals 10, each client has an IID subset of the entire dataset. When $N_c = 2$ or 5, the samples on each client are divided by the labels.

Figure 2.15 illustrates the convergence curves of the three federated learning algorithms in terms of the communication rounds, communication costs and test accuracy. We can see that both T-FedAvg and FedAvg are better than CMFL in terms of convergence, and T-FedAvg outperforms FedAvg in late stages. Meanwhile, T-FedAvg requires the least amount of communication to achieve the same accuracy, verifying the conclusion drawn from Table 2.11.

Table 2.11 Test accuracies achieved over Non-IID data for different N_c values

Accuracy (%)	MNIST		CIFAR10 (ResNet*)	
	$N_c = 2$	$N_c = 5$	$N_c = 2$	$N_c = 5$
FedAvg	82.61 ± 4.31	89.24 ± 0.38	40.17 ± 5.6	71.47 ± 0.23
CMFL	81.51 ± 2.29	88.30 ± 0.52	39.30 ± 6.1	70.26 ± 0.63
T-FedAvg	87.29 ± 0.89	90.04 ± 0.68	40.46 ± 4.3	76.69 ± 0.67

The better performance of T-FedAvg observed on the CIFAR10 dataset seems counterintuitive. However, as demonstrated in Sect. 2.4.4, if Assumption 2.1 holds, quantization can reduce the weight divergence in federated learning, resulting in more robust classification performance on Non-IID data. The results in Fig. 2.15 also confirm the performance improvement of T-FedAvg compared to FedAvg.

It should be noted that the poor performance of federated learning on Non-IID data remains a challenging problem [41]. Because in this case, the local stochastic gradient cannot be regarded as an unbiased estimation of the global gradient [40]. Theoretically, T-FedAvg reduces the upstream and downstream communication, and hence we can increase the number of communication rounds under the same budget to alleviate performance degradation.

2.4.5.4 Results on Unbalanced Data

All the above experiments are performed with the dataset being equally divided. This section mainly studies the performance of the proposed algorithm with unbalanced client dataset sizes [47]. Specifically, let $S_K = \{|\mathcal{D}_1|, |\mathcal{D}_2|, ..., |\mathcal{D}_K|\}$ represent the union of samples on K clients, and the degree of imbalance is defined by the ratio β:

$$\beta = \frac{\text{median}\{S_K\}}{\max\{S_K\}}, \tag{2.53}$$

where the median operation used in the formula helps to accommodate long-tailed distributions and possible outliers [48].

When $\beta = 0.1$, most of the samples are stored on a few clients, while when $\beta = 1$, almost all clients store the same number of samples. The simulation experiments are conducted using the MNIST dataset with a fixed participation ratio of 0.3. The test accuracies of FedAvg and T-FedAvg under different β are shown in Fig. 2.16.

It can be seen that the effect of imbalance on the performance of the federated learning algorithm is not obvious when data is IID.

Fig. 2.16 Test accuracies achieved by the MLP on MNIST after 100 rounds of iterations with FedAvg, CMFL and T-FedAvg under unbalanced data, where the number of clients and participation ratio are set to 100 and 0.1, respectively

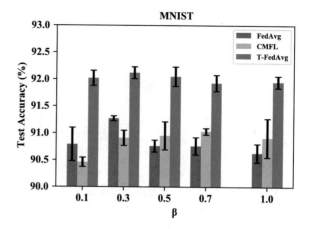

2.4.5.5   Influence of Participation Ratio

Because of MLP's robustness to Non-IID data, and to minimize the model's influence, this section solely employs MLP to conduct tests on the MNIST dataset. Figure 2.17 depicts T-FedAvg's test accuracy on IID and Non-IID MNIST for various participation ratios *lambda*, as well as the upstream communication costs.

It can be seen that T-FedAvg is relatively robust to changes in participation ratios for IID and Non-IID data, and the performance fluctuations decrease as λ increases. In general, reducing the value of λ usually has a negative impact on the learning rate and the final accuracy achieved within a fixed epoch, and the negative impact is more pronounced on the non-IID data (refer to Fig. 2.17). However, once λ increases to a certain value, the performance gain becomes less significant and sometimes even starts to drop.

Intuitively, the upstream and downstream communication costs increase as λ increases. And compared with FedAvg, the more clients when the participation ratio is fixed, the more communication cost T-FedAvg can save. This is important in practical applications because the number of clients in the real world is usually very large.

2.4.5.6   Computational Complexity

In this section, we first analyze the computational complexity of FedAvg and T-FedAvg, and then conduct empirical experiments to compare the running time of FedAvg, T-FedAvg and CMFL.

As stated in Mu et al. [49], for one mini-batch training on each client, the time complexity of FedAvg is $O(BD)$, where B is the batch size and D is the number of elements in w. Correspondingly, the time complexity of T-FedAvg in each iteration can be divided into three parts: $O(\frac{D}{2} + 1)$ for calculating α^q by the Eq. (2.22).

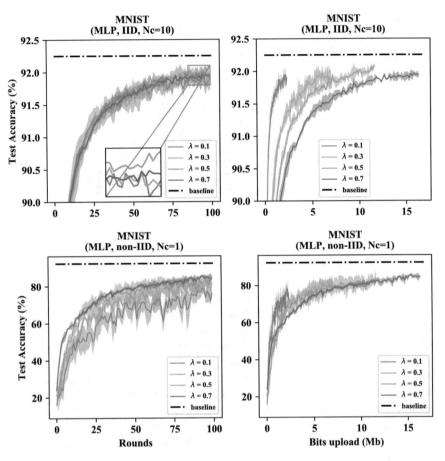

Fig. 2.17 Test accuracies achieved by T-FedAvg and uploaded bits when training MLP on MNIST with IID and non-IID distribution in fixed rounds at different participation ratios $(0.1, 0.3, 0.5, 0.7)$

$O(BD)$ for differentiating w by Eq. (2.23). And $O(D)$ for quantization. Therefore, the overall time complexity of T-FedAvg per iteration per client is $O((B + \frac{3}{2})D)$. For the server, the average time complexity of FedAvg and T-FedAvg is $O((\lambda K - 1)D)$ and $O(\lambda K D)$, respectively, where λ is the participation ratio, K is the total number of clients, $\lambda K \geq 1$.

Now we investigate the runtime of different federated learning algorithms. The PyTorch framework [30] is used to implement three methods on a workstation with Intel Xeon(R) CPU E5-2620 v4 @ 2.10GHz × 32 and 1 NVIDIA GTX1080Ti GPU, and the system version is Ubuntu 16.04 LTS. The runtime of each client for 5 local epochs is counted, the results obtained in 20 independent runs are presented in Table 2.12.

As we can see, when training MLP, the elapsed time of T-FedAvg is almost the same as FedAvg and is much less than CMFL. However, as the depth of global model

Table 2.12 Elapsed time for each client to complete five local epochs over different federated learning algorithms

Time (s)	MNIST	CIFAR10	
	MLP	CNN	ResNet*
FedAvg	1.06 ± 0.03	7.92 ± 0.34	21.74 ± 0.38
CMFL	1.62 ± 0.06	12.48 ± 0.18	27.68 ± 0.35
T-FedAvg	1.06 ± 0.10	8.97 ± 0.23	38.22 ± 0.50

increases, the time complexity of our algorithm will increase significantly. Indeed, this will be a potential limitation of model quantization when the model becomes deeper, despite that the proposed method has already reduced a large number of quantization factors. In the future, we will work out solutions to reduce the time complexity of T-FedAvg, for example, by implementing quantization only on the last local epoch.

2.5 Summary

Federated learning, as the mainstream method to protect the security and privacy of distributed machine learning, faces the challenges of communication bottleneck and Non-IID data distributions in real-world applications. To solve the above problems, this chapter introduces layer-wise asynchronous update and model quantization into federated learning systems. Specifically, we first design a layer-wise asynchronous model update approach to reduce the number of the parameters to be transmitted. Secondly, a model quantization algorithm is used to simplify the local models for upstream communication reduction, and a ternary aggregation method is proposed to reduce downstream communication cost. The satisfactory performance proves the efficiency of the proposed algorithms.

References

1. speedtest.net: United kingdom mobile speedtest data (2016). https://www.speedtest.net/reports/united-kingdom/
2. Sze, V., Chen, Y.H., Yang, T.J., Emer, J.S.: Efficient processing of deep neural networks: a tutorial and survey. Proc. IEEE **105**(12), 2295–2329 (2017)
3. Chen, Y., Sun, X., Jin, Y.: Communication-efficient federated deep learning with layerwise asynchronous model update and temporally weighted aggregation. IEEE Trans. Neural Netw. Learn. Syst. **31**(10), 4229–4238 (2019)
4. McMahan, B., Moore, E., Ramage, D., Hampson, S., Arcas, B.A.y.: Communication-Efficient Learning of Deep Networks from Decentralized Data. In: Singh, A., Zhu, J. (eds.) Proceedings of the 20th International Conference on Artificial Intelligence and Statistics, Proceedings of Machine Learning Research, vol. 54, pp. 1273–1282. PMLR (2017). https://proceedings.mlr.press/v54/mcmahan17a.html

5. Kairouz, P., McMahan, H.B., Avent, B., Bellet, A., Bennis, M., Bhagoji, A.N., Bonawitz, K., Charles, Z., Cormode, G., Cummings, R., et al.: Advances and open problems in federated learning (2019). arXiv:1912.04977
6. Moreno-Torres, J.G., Raeder, T., Alaiz-Rodríguez, R., Chawla, N.V., Herrera, F.: A unifying view on dataset shift in classification. Pattern Recognit. **45**(1), 521–530 (2012)
7. Quiñonero-Candela, J., Sugiyama, M., Lawrence, N.D., Schwaighofer, A.: Dataset Shift in Machine Learning. MIT Press (2009)
8. Eichner, H., Koren, T., McMahan, B., Srebro, N., Talwar, K.: Semi-cyclic stochastic gradient descent. In: International Conference on Machine Learning, pp. 1764–1773. PMLR (2019)
9. Konečný, J., McMahan, H.B., Yu, F.X., Richtárik, P., Suresh, A.T., Bacon, D.: Federated learning: Strategies for improving communication efficiency (2016). arXiv:1610.05492
10. Acharya, J., Canonne, C.L., Tyagi, H.: Inference under information constraints i: lower bounds from chi-square contraction. IEEE Trans. Inf. Theory **66**(12), 7835–7855 (2020)
11. Zhang, Y., Duchi, J.C., Jordan, M.I., Wainwright, M.J.: Information-theoretic lower bounds for distributed statistical estimation with communication constraints. In: NIPS, pp. 2328–2336. Citeseer (2013)
12. Han, Y., Özgür, A., Weissman, T.: Geometric lower bounds for distributed parameter estimation under communication constraints. In: Conference On Learning Theory, pp. 3163–3188. PMLR (2018)
13. Braverman, M., Garg, A., Ma, T., Nguyen, H.L., Woodruff, D.P.: Communication lower bounds for statistical estimation problems via a distributed data processing inequality. In: Proceedings of the Forty-Eighth Annual ACM Symposium on Theory of Computing, pp. 1011–1020 (2016)
14. Suresh, A.T., Felix, X.Y., Kumar, S., McMahan, H.B.: Distributed mean estimation with limited communication. In: International Conference on Machine Learning, pp. 3329–3337. PMLR (2017)
15. Chraibi, S., Khaled, A., Kovalev, D., Richtárik, P., Salim, A., Takáč, M.: Distributed fixed point methods with compressed iterates (2019). arXiv:1912.09925
16. Caldas, S., Konečny, J., McMahan, H.B., Talwalkar, A.: Expanding the reach of federated learning by reducing client resource requirements (2018). arXiv:1812.07210
17. Hamer, J., Mohri, M., Suresh, A.T.: Fedboost: A communication-efficient algorithm for federated learning. In: International Conference on Machine Learning, pp. 3973–3983. PMLR (2020)
18. Park, J., Samarakoon, S., Bennis, M., Debbah, M.: Wireless network intelligence at the edge (2018). arXiv:1812.02858
19. Samarakoon, S., Bennis, M., Saad, W., Debbah, M.: Federated learning for ultra-reliable low-latency v2v communications. In: 2018 IEEE Global Communications Conference (GLOBE-COM), pp. 1–7. IEEE (2018)
20. Abari, O., Rahul, H., Katabi, D.: Over-the-air function computation in sensor networks (2016). arXiv:1612.02307
21. Jeong, E., Oh, S., Kim, H., Park, J., Bennis, M., Kim, S.L.: Communication-efficient on-device machine learning: federated distillation and augmentation under non-iid private data (2018). arXiv:1811.11479
22. Ahn, J.H., Simeone, O., Kang, J.: Wireless federated distillation for distributed edge learning with heterogeneous data. In: 2019 IEEE 30th Annual International Symposium on Personal, Indoor and Mobile Radio Communications (PIMRC), pp. 1–6. IEEE (2019)
23. Krizhevsky, A.: One weird trick for parallelizing convolutional neural networks (2014). arXiv:1404.5997
24. Yosinski, J., Clune, J., Bengio, Y., Lipson, H.: How transferable are features in deep neural networks? In: Advances in Neural Information Processing Systems, pp. 3320–3328 (2014)
25. Krizhevsky, A., Sutskever, I., Hinton, G.E.: Imagenet classification with deep convolutional neural networks. Adv. Neural Inf. Process. Syst. **25**, 1097–1105 (2012)
26. LeCun, Y., Bottou, L., Bengio, Y., Haffner, P.: Gradient-based learning applied to document recognition. Proc. IEEE **86**(11), 2278–2324 (1998)

27. Anguita, D., Ghio, A., Oneto, L., Parra, X., Reyes-Ortiz, J.L.: A public domain dataset for human activity recognition using smartphones. In: ESANN (2013)
28. Gers, F.A., Schmidhuber, J., Cummins, F.: Learning to Forget: Continual Prediction With lstm (1999)
29. Jacob, B., Kligys, S., Chen, B., Zhu, M., Tang, M., Howard, A., Adam, H., Kalenichenko, D.: Quantization and training of neural networks for efficient integer-arithmetic-only inference. In: Proceedings of the IEEE Conference on Computer Vision and Pattern Recognition, pp. 2704–2713 (2018)
30. Paszke, A., Gross, S., Massa, F., Lerer, A., Bradbury, J., Chanan, G., Killeen, T., Lin, Z., Gimelshein, N., Antiga, L., Desmaison, A., Kopf, A., Yang, E., DeVito, Z., Raison, M., Tejani, A., Chilamkurthy, S., Steiner, B., Fang, L., Bai, J., Chintala, S.: Pytorch: an imperative style, high-performance deep learning library. In: Advances in Neural Information Processing Systems, pp. 8024–8035 (2019)
31. Dettmers, T.: 8-bit approximations for parallelism in deep learning (2015). arXiv:1511.04561
32. Courbariaux, M., Bengio, Y., David, J.P.: Low precision storage for deep learning (2014). arXiv:1412.7024
33. Li, F., Zhang, B., Liu, B.: Ternary weight networks. In: Proceedings of Conference on Neural Information Processing Systems: Workshop (2016)
34. Zhu, C., Han, S., Mao, H., Dally, W.J.: Trained ternary quantization. In: Proceedings of International Conference on Learning Representations (2017)
35. Polino, A., Pascanu, R., Alistarh, D.: Model compression via distillation and quantization (2018). arXiv:1802.05668, https://arxiv.org/pdf/1802.05668.pdf
36. Gray, R.M., Neuhoff, D.L.: Quantization. IEEE Trans. Inf. Theory **44**(6), 2325–2383 (1998)
37. Li, X., Huang, K., Yang, W., Wang, S., Zhang, Z.: On the convergence of fedavg on non-iid data. In: Proceedings of International Conference on Learning Representations (2020)
38. Stich, S.U.: Local sgd converges fast and communicates little. In: Proceedings of International Conference on Learning Representations (2019)
39. Qu, Z., Lin, K., Kalagnanam, J., Li, Z., Zhou, J., Zhou, Z.: Federated learning's blessing: Fedavg has linear speedup (2020). arXiv:2007.05690, https://arxiv.org/pdf/2007.05690.pdf
40. Zhao, Y., Li, M., Lai, L., Suda, N., Civin, D., Chandra, V.: Federated learning with non-iid data (2018). arXiv:1806.00582, https://arxiv.org/pdf/1806.00582.pdf
41. McMahan, B., Moore, E., Ramage, D., Hampson, S., Arcas, B.A.y: Communication-efficient learning of deep networks from decentralized data. In: Proceedings of Artificial Intelligence and Statistics, pp. 1273–1282. PMLR (2017)
42. Wang, L., Wang, W., Li, B.: CMFL: Mitigating communication overhead for federated learning. In: Proceedings of International Conference on Distributed Computing Systems, pp. 954–964. IEEE (2019)
43. He, K., Zhang, X., Ren, S., Sun, J.: Deep residual learning for image recognition. In: Proceedings of the IEEE Conference on Computer Vision and Pattern Recognition, pp. 770–778 (2016)
44. Lecun, Y., Bottou, L., Bengio, Y., Haffner, P.: Gradient-based learning applied to document recognition. Proc. IEEE **86**(11), 2278–2324 (1998)
45. Krizhevsky, A., Hinton, G.: Learning multiple layers of features from tiny images. Technical report, Dept. Comput. Sci., Univ. Toronto, Toronto, ON, Canada (2009)
46. Wang, H., Yurochkin, M., Sun, Y., Papailiopoulos, D., Khazaeni, Y.: Federated learning with matched averaging. In: International Conference on Learning Representations (2020). https://openreview.net/forum?id=BkluqlSFDS
47. Sattler, F., Wiedemann, S., Müller, K.R., Samek, W.: Robust and communication-efficient federated learning from non-iid data. IEEE Trans. Neural Netw. Learn. Syst. **31**(9), 3400–3413 (2019)
48. Bowman, A.W., Azzalini, A.: Applied Smoothing Techniques for Data Analysis: The Kernel Approach with S-Plus Illustrations, vol. 18. OUP Oxford (1997)
49. Mu, Y., Liu, W., Liu, X., Fan, W.: Stochastic gradient made stable: A manifold propagation approach for large-scale optimization. IEEE Trans. Knowl. Data Eng. **29**(2), 458–471 (2016)

Chapter 3
Evolutionary Multi-objective Federated Learning

Abstract Different from model quantization and partial model uploads presented in the previous chapter, evolutionary federated learning, more specifically, evolutionary federated neural architecture search, aims to optimize the architecture of neural network models, thereby reducing the communication costs caused by frequent model transmissions, generating lightweight neural models that are better suited for mobile and other edge devices, and also enhancing the final global model performance. To achieve this, scalable and efficient encoding methods for deep neural architectures must be designed and evolved using multi-objective evolutionary algorithms. This chapter presents two multi-objective evolutionary algorithms for federated neural architecture search. The first one employs a probabilistic representation of deep neural architectures that describes the connectivity between two neighboring layers and simultaneously maximizing the performance and minimizing the complexity of the neural architectures using a multi-objective evolutionary algorithm. However, this evolutionary framework is not practical for real-time optimization of the neural architectures in a federated environment. To tackle this challenge, a real-time federated evolutionary neural architecture search is then introduced. In addition to adopting a different neural search space, a double sampling strategy, including sampling subnetworks from a pretrained supernet and sampling clients for model update, is proposed so that the performance of the neural architectures becomes more stable, and each client needs to train one local model in one communication round, thereby preventing sudden performance drops during the optimization and avoiding training multiple submodels in one communication round. This way, evolutionary neural architecture search is made practical for real-time real-world applications.

3.1 Motivations and Challenges

Federated learning (FL) is an emerging technology proposed for privacy preservation, when the central cloud is likely to train a machine learning (ML) model from distributed client devices. However, as introduced previously, FL consumes much more communication resources than the standard centralized ML, since model gradients or parameters need to be downloaded and uploaded frequently between the central server and connected clients during the training period. And this issue has become a bottleneck for real-world applications of FL.

Using multi-objective evolutionary algorithms (MOEAs) to reduce the model complexity is an effective approach to mitigating the aforementioned communication problems, since MOEAs are able to optimize the structure of shared global model in FL to reduce not only the model size but also the learning errors. And then the communication costs will be decreased by transmitting small-size models between the server and clients. Note, however, that most MOEA-based approaches are designated for *offline* optimization, in which model structure optimization and parameter training are separately performed at each generation.

Offline evolutionary approaches to learning suffer from two major drawbacks. First, they are not well suited for real-time FL systems such as online recommenders, because all new candidate neural architectures often need to be re-initialized and re-trained from scratch, which not only degrades the online model performance but also increases the computational costs. Second, since MOEAs are population-based search methods, multiple models need to be trained within one communication round, increasing both communication cost and computational burdens. Finally, once the evolutionary search is completed, the found models must be retrained from scratch, consuming additional communication resources in FL systems. Therefore, it is necessary to develop an efficient real-time evolutionary FL framework to support online optimization requirements.

In the next section, we first introduce the classic offline multi-objective evolutionary FL method in detail, followed by a real-time evolutionary neural architecture search scheme optimizing the performance, model complexity and computational complexity.

3.2 Offline Evolutionary Multi-objective Federated Learning

Offline evolutionary multi-objective federated learning belongs to Pareto-based multi-objective machine learning [1] and it differs from the centralized multi-objective learning only in the model training part in fitness evaluations. That is, the conventional Pareto-based multi-objective machine learning adopts a centralized approach to model training, while offline evolutionary multi-objective federated learning trains the models in a distributed way. In the following, we will introduce the algorithm details of the offline evolutionary multi-objective federated learning approach from three aspects. At first, a sparse evolutionary network encoding method called the modified SET algorithm will be introduced. This is followed by a description of the evolutionary multi-objective network search. Finally, the overall framework is presented.

3.2.1 Sparse Network Encoding with a Random Graph

One common method for encoding the connection of neural networks in evolutionary algorithm is the direct binary encoding of all possible connections between neurons [1]. The basic idea is to construct a large binary connection matrix, in which '1' represents there is a connection between two nodes and '0' means no connection. The advantage of this approach is that the encoded chromosome can be directly used for crossover and mutation; however, this method is not scalable to multi-layer neural networks with a large number of neurons. Therefore, a modified sparse evolutionary training (SET) [2] scheme is proposed to enhance both scalability and flexibility in representing sparse neural networks.

The SET representation does not encode the connections of neural networks and then performs crossover or mutation, like in the conventional evolutionary methods. Instead, SET utilizes the Erdos Rènyi random graph [3] to determine the connection probability between the nodes of every two neighbouring neural network layer. And the connection probability is defined in Eq. (3.1).

$$
\begin{aligned}
p(W_{ij}^k) &= \frac{\varepsilon(n^k + n^{k-1})}{n^k n^{k-1}}, \\
n^W &= n^k n^{k-1} p(W_{ij}^k),
\end{aligned}
\tag{3.1}
$$

where W_{ij}^k represents a connection weight between nodes i and j within the kth sparse weight matrix W^k, n^k and n^{k-1} are the number of neurons in layers k and $k-1$, respectively, and ε is a SET hyperparameter for controlling sparsity, and n^W is the total number of connections between layers k and $k-1$. Apparently, the larger the number of neurons n^k and n^{k-1} is, the more sparse the connection weight matrix W^k will be.

The original SET algorithm suggests that ξ fraction of the smallest connection weights be removed after each training epoch to mitigate the side-effect caused by the randomly initialized graph. However, removing the least important model parameters frequently may bring in unexpected performance fluctuations during training, which turns out to be more serious in FL. To resolve this issue, a modified SET algorithm is proposed by conducting the removing operation at the last training epoch only to stabilize the overall training process. And a sparsely connected neural network can be further evolved, resulting in much fewer parameters to be downloaded or uploaded. The pseudo code of the modified SET is listed in Algorithm 1.

3.2.2 Evolutionary Multi-objective Neural Architecture Search

In order to use multi-objective evolutionary algorithms to optimize the structure of neural networks, the first step is to reformulate federated learning as a two-objective

---

Algorithm 1 Modified SET algorithm

1: Set ε and ξ
2: **for** each fully-connected layer of the neural network **do**
3:     Replace weight matrices by Erdos Rènyi random graphs given by ε in Eq. (3.1)
4: **end for**
5: Initialize weights
6: Start training
7: **for** each training epoch **do**
8:     Training and updating corresponding weights
9: **end for**
10: **for** each weight matrix **do**
11:     Remove a fraction ξ of the smallest $|weights|$
12: **end for**

---

optimization problem [4]. One objective is usually the final model performance like the validation error E_t of the global model, and the other one is the model size Ω_t at the tth communication round. To minimize these two objectives, both hyperparameters as well as the weight connections of neural network models are evolved. The hyperparameters consist of the number of hidden layers, the number of neurons in each hidden layer, and the learning rate η. And the weight connections are evolved using the modified SET introduced in Algorithm 1, which contains two hyperparameters ε and ξ used to determine the sparsity of network connections.

Consequently, two types of decision variables appear in the chromosome of the evolutionary algorithm, namely integers encoded by binary numbers and real numbers. For example, the number of hidden layers and the number of neurons in each layer are both integers that can be converted into binary numbers, while the real-valued parameters like the learning rate and SET variables remain to be real values. Figure 3.1 provides an example of a chromosome and the corresponding encoded multi-layer perceptron (MLP) neural network. It can be seen that both real numbers and integers are encoded into one chromosome, where the decision variables representing the number of hidden layers are queued at the last to allow different individuals to have chromosome of different lengths. Note that the decoded values for the number of hidden layers will be increased by one to make sure that there is at least one neuron in a hidden layer.

A similar encoding method is also applied to the convolutional neural network (CNN) with minor differences due to different model structures. A typical CNN contains a number of convolutional layers followed by a few fully connected layers. In addition to the number of convolutional layers and channels encoded by integers, a randomly chosen integer between 3 and 5 is adopted to encode the kernel size. An illustrative example is shown in Fig. 3.2.

Each encoded chromosome represents one sparsely connected neural network, which, when decoded, is the global model trained in federated learning to calculate the validation accuracy A_t. After that, the global validation error E_t can be further computed as one fitness value of the two objectives in evolutionary optimization.

Fig. 3.1 An encoded MLP
neural network and its
corresponding chromosome

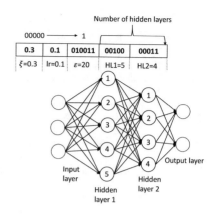

Fig. 3.2 An encoded CNN
and its corresponding
chromosome

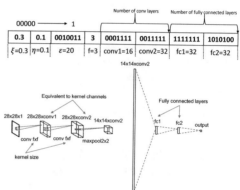

The other objective, the model complexity Ω_t, can be easily derived by averaging
the number of model weights uploaded from all the clients (Eq. (3.2)).

$$E_t = 1 - A_t,$$

$$\Omega_t = \sum_{k=1}^{K} \Omega_k / K, \qquad (3.2)$$

where K is the number of connected clients, and ω_i indicates the number of parameters of the kth client model. After combining the modified SET representation introduced in Sect. 3.2.1 with the standard FedAvg algorithm [5], the fitness evaluation
part of the evolutionary optimization is described in Algorithm 2.

In the algorithm, K indicates the total numbers of clients, B is the local mini-
batch size, η is the learning rate, Ω represents the number of model parameters, i
is one individual solution representing a particular neural network with a modified
SET topology as a global model used in FedAvg, and R is the total population. Once
the structure and connectivity of the neural network models are determined by the
decoded chromosomes in the evolutionary algorithm, the weights or parameters will

---

Algorithm 2 The modified SET FedAvg optimization

1: **for** each population $i \in R$ **do**
2:     Globally initialize θ_t^i with a Erdos Rènyi random graph
3:     **for** each communication round $t = 1, 2, \ldots$ **do**
4:         Select $m = C \times K$ clients, $C \in (0, 1)$ clients
5:         $\Omega_t = 0$
6:         **for** each client $k \in m$ **do**
7:             **for** each local epoch e from 1 to E **do**
8:                 **for** batch $b \in B$ **do**
9:                     $\theta_e^k = \theta_t^i - \eta \nabla \ell(\theta_t^i; b)$
10:                 **end for**
11:                 remove a fraction of ξ smallest values in θ^k
12:             **end for**
13:             $\theta_{t+1}^i = \theta_t^i + \frac{n_k}{n} \theta^k$
14:             $\Omega^k = f(\theta^k)$ (calculate the number of model parameters)
15:             $\Omega_t = \Omega_t + \frac{n_k}{n} \Omega^k$
16:         **end for**
17:     **end for**
18:     Evaluate model accuracy through θ^i
19:     Calculate model error as objective one f_i^1
20:     Set Ω_t as objective two f_i^2
21: **end for**
22: return f^1 and f^2

---

be trained by the mini-batch SGD in a distributed environment until the specified maximum number of communication round is reached.

3.2.3 Overall Framework

The overall framework of the bi-objective evolutionary optimization of federated learning using the elitist non-dominated sorting genetic algorithm (NSGA-II) [6] is shown in Fig. 3.3. NSGA-II is a widely used multi-objective evolutionary algorithm aiming at achieving a set of diverse Pareto optimal solutions (in practice non-dominated solutions) by comparing the dominance relationships and the crowding distance of the solutions in the population. And the pseudo-code is summarized in Algorithm 3.

In the algorithm, M parent solutions P_t are randomly initialized with binary and real-valued chromosomes at the first generation. And then each pair of parents are chosen to generate two offspring using crossover and mutation. Note that the encoded chromosomes are mixed up with binary and real numbers. Therefore, two types of operators are adopted here, namely one-point crossover and flip mutation for the binary decision variables, and the simulated binary crossover (SBX) and polynomial mutation [7] for the real decision variables. After M offspring are generated, both

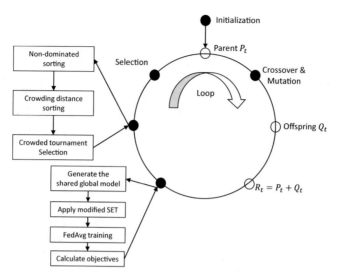

Fig. 3.3 A framework for multi-objective federated neural architecture search using NSGA-II

Algorithm 3 Multi-objective evolutionary optimization
___
1: Randomly generate parent solutions P_t where $|P_t| = M$
2: **for** each generation $t = 1, 2, ...$ **do**
3:     Generate offspring $|Q_t| = M$ through crossover and mutation
4:     $R_t = P_t + Q_t$
5:     Evaluate f_t^1 and f_t^2 by Algorithm 5
6:     $f \leftarrow (f_t^1, f_t^2)$
7:     **for** each solution in R_t **do**
8:         Perform non-dominated sorting and calculate the crowding distance on f
9:         Select high-ranking solutions from R_t
10:         Let $P_t = R_t$
11:     **end for**
12: **end for**
___

parents and offspring are combined to get the whole population with $2M$ solutions, all of which are subject to environmental selection.

Except for the first evolutionary generation, only the objective values of offspring are calculated, since the objective values of the parents have already been computed at the previous generation. And then, the objective values of offspring are integrated with those of parents and sorted based on the non-dominance relationships and the crowding distance. Finally, M highest ranking individuals are selected to be the new parents for the next generation. This process repeats until the evolutionary optimization algorithm converges, and a set of non-dominated solutions will be found, each representing a Pareto optimal neural architecture.

3.2.4 Empirical Results

In this section, we are going to present empirical results to test the performance of the proposed algorithm. Two sets of experiments are conducted, one for comparing the model performance with and without applying the modified SET algorithm in federated learning, and the other for comparing two objective values among a set of found Pareto optimal solutions obtained by the proposed multi-objective evolutionary federated learning and the standard fully connected neural networks on the Non-IID data.

And two popular neural network models, the MLP neural network and the CNN are considered in our experiments, both trained and validated on a benchmark dataset MNIST [8]. The standard MLP neural network follows the model structure used in [5] that contains two hidden layers, each having 200 neurons with a total of 199210 model parameters. For the CNN model, two convolutional layers adopt 3×3 kernel filters with 32 and 64 channels, respectively, followed by a 2×2 max pooling layer, and two fully connected layers with 128 and 10 neurons, respectively, resulting in a total of 1625866 parameters. Besides, the conventional mini-batch SGD algorithm is adopted for local model training with the learning rate of 0.1 and a mini-batch size of 50. These can be regarded as the baseline model structure used in our experiments.

The federated learning settings are also listed here. The total number of clients K are set to be 100. We assume that all clients are connected at each communication round. For local client training, the number of local epochs is set to be 5. For Non-IID data, the dataset is sorted according to the labeled class and evenly divided into 200 shards. And two randomly selected shards are allocated to each client without replacement. Thus, one client device only has two classes of the training data.

Besides, the population size is set to be 20, and the maximum number of generations is 50. For the binary number of the encoded chromosome, the one-point crossover probability is 0.9 and bit-flip mutation probability is 0.1. Similarly, the SBX crossover is applied on the real values with a probability of 0.9 and $n_c = 2$, and the polynomial mutation with a probability of 0.1 and $n_m = 20$. In addition, the communication rounds required to compute the model accuracy for fitness evaluation is set to be 10. The two SET hyperparameters, ε and ξ, are used to control the sparsity.

3.2.4.1 Influence of the Neural Network Sparsity on the Performance

In the first set of the experiments, the influence of different sparsity levels on the model performance is examined by using different SET hyperparameters. And three ε values (100, 50, 20) and two ξ values (0, 0.3) are selected for both MLPs and CNNs with the aforementioned standard structure. Besides, the communication rounds used in the modified SET FedAvg algorithm is 500 for the MLPs and 200 for the CNNs. The results are shown in Fig. 3.4.

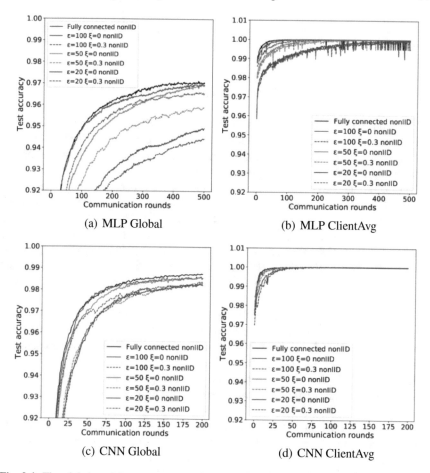

Fig. 3.4 The global model test accuracy and the average client accuracy for MLPs and CNNs, respectively

At first, the convergence properties of the global model are discussed on both the central server and clients. The average client accuracy is computed by averaging the local test accuracy over all connected clients. It can be seen that the average client accuracy converges to nearly 100% within approximately 200 rounds for the MLPs and 50 rounds for the CNNs, respectively, refer to Fig. 3.4(b)(d). On the other hand, the global test accuracy on the server side converges much slower as shown in Fig. 3.4(a)(c), which indicates that learning on the server is more challenging. To take a closer look at the global model performance, it is obvious to see that the global test accuracy degrades when the sparsity level of neural network models increase.

Overall, the test accuracy of the global model tends to decline when we tune the SET parameters to raise the sparseness of the shared neural network model in our experiments. In other words, using the modified SET FedAvg algorithm only cannot maximize the global learning accuracy and minimize the communication costs at the same time.

3.2.4.2   Evolved Federated Learning Models

In the second part of our experiments, we adopt NSGA-II to get a set of Pareto optimal neural networks at the last evolutionary generation. Both two objective values of the global model performance and model complexity between the searched solutions and the original fully connected one will be compared.

The communication round used for computing model accuracy is set to be 10. The maximum number of hidden layers for the MLPs is 4, and the maximum number of neurons per layer is set to be 256. And for the CNNs, the maximum number of convolutional layers and fully connected layers are both set to be 3, and the kernel size is either 3 or 5. Moreover, the range of the learning rate is between 0.01 and 0.3, and ξ ranges from 0.01 to 0.55. The maximum value of binary encoded ε is 128. All these model hyperparameters are encoded into the chromosome of the individuals in the population.

The Pareto optimal solutions of the last evolutionary generation are presented in Fig. 3.5, and two types of non dominated solutions are selected for further discussions, those with a low global model error (high accuracy), and those near knee point of the frontier, as suggested in [1]. Therefore, refer to Fig. 3.5, two high-accuracy solutions (High1 and High2) and two solutions near the knee point (Knee1 and Knee2) are used to compare their performance with the standard fully connected neural network models.

The selected solutions from the frontier and the standard model are both trained from scratch and all the results are listed in Tables 3.1 and 3.2. Moreover, the global test accuracy over communication rounds are also plotted in Fig. 3.6.

For the MLPs, it is obvious to find out that solution High1 has a global test accuracy of 97.32% on Non-IID data, which is 0.28% higher than that of the standard fully

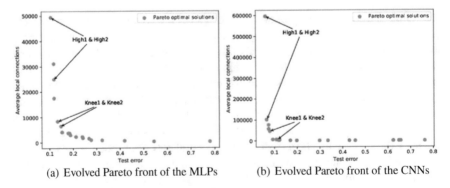

(a) Evolved Pareto front of the MLPs   (b) Evolved Pareto front of the CNNs

Fig. 3.5 Pareto front, High1, High2, Knee1, Knee2 are selected for validation

Table 3.1 Hyper-parameters and validation results for the MLPs

Parameters	Knee1	Knee2	High1	High2	Standard
Hidden layer1	49	53	86	109	200
Hidden layer2	/	/	/	/	200
ε	10	8	66	34	/
ξ	0.1106	0.0764	0.1106	0.1566	/
Learning rate η	0.3	0.2961	0.3	0.3	0.1
Test accuracy nonIID (%)	94.85	94.88	97.32	96.21	97.04
Connections nonIID	8,086	6,143	45,530	24,055	199,210

Table 3.2 Hyper-parameters and validation results for the CNNs

Parameters	Knee1	Knee2	High1	High2	Standard
Conv layer1	17	5	53	33	32
Conv layer2	/	/	/	/	64
Fully connected layer1	29	21	208	31	128
Fully connected layer2	/	/	/	/	/
Kernel size	5	5	5	5	3
ε	18	8	66	20	/
ξ	0.1451	0.1892	0.0786	0.1354	/
Learning rate η	0.2519	0.2388	0.2776	0.2503	0.1
Test accuracy nonIID (%)	97.92	97.7	98.52	98.46	98.75
Connections nonIID	39457	6804	553402	90081	1,625,866

connected model, with only 12.08% of 199210 connections. Knee1 and Knee2 have a similar global test accuracy with 94.85% and 94.88%, respectively. Although their model accuracy is around 2% lower than that of the standard MLP, they only have approximately 3.5% weight connections of the fully connected one.

Similar observations can be made on the two high-accuracy solutions for CNNs. The global test accuracy of High1 and High2 are 98.52% and 98.46%, respectively, both of which are about 0.2% lower than the standard CNN. However, High1 and High2 only have about 10% of the connections of the fully connected CNN. On the other hand, Knee1 and Knee2 also have great global model performance of 97.92

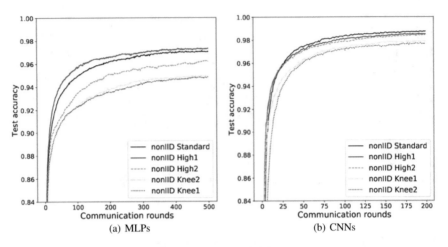

Fig. 3.6 The global test accuracy for both MLPs and CNNs

and 97.7% test accuracy, respectively, both of which only have approximately 1.2% of weight connections compared to the standard fully connected one.

To conclude, the proposed algorithm can achieve a set of Pareto optimal solutions, which is able to reduce the number of parameters in the shared model without deteriorating the model performance.

3.3 Real-Time Evolutionary Federated Neural Architecture Search

The aforementioned multi-objective evolutionary federated learning algorithm [9] is inherently an offline evolutionary optimization framework. Just like offline evolutionary neural architecture search (NAS) methods proposed in [10–12], the model parameters of the offspring at each generation should be randomly initialized and trained from scratch, before evaluating the validation accuracy of offspring. This approach brings about two shortcomings. Firstly, it requires a large amount of computational resources; Secondly, the reinitialized models may have significant performance drop, making them inappropriate for instant usage. What is worse, re-training optimized models in federated learning must consume extra communication costs.

Therefore, it is highly desirable to develop an online or real-time federated evolutionary NAS framework, which is called RT-FedEvoNAS. Unlike data-driven evolutionary optimization [13], the term 'online' used in federated evolutionary NAS means that the neural networks are already in use during the evolutionary neural architecture search. Consequently, the offspring model performance is not allowed to drop dramatically. In addition, the searched neural network models should be well trained at the last evolutionary generation to avoid consuming extra communication resources.

In this section, we will introduce the supernet based neural architecture encoding method used in federated learning at first. This is followed by a description of the double sampling techniques based on the global supernet model. Finally, the overall framework will be presented together with empirical studies.

3.3.1   Network Architecture Encoding Based On Supernet

In addition to the data privacy issue, communication costs and the local computational overheads are always the primary concerns in federated learning. So a light-weight supernet structure is implemented as the global model on the server. As shown in Fig. 3.7, the global supernet includes a convolutional block, 3x4 choice blocks, and a fully connected layer with a maximum of 26 hidden layers. To be more specifically, the convolutional block is composed of three sequentially connected layers with a convolutional layer, a batch normalization layer [14], and a rectified linear unit (ReLU) layer. And a choice block contains four branches, each of which represents a pre-defined candidate block, namely identity block, residual block, inverted residual block and depthwise separable block, as shown in Fig 3.8. These candidate blocks are categorized into two groups, one is called the normal block, whose input and output share the same channel dimension, whereas the other is called the reduction block, whose output channel dimension is doubled and the spatial dimension is quartered compared to its input.

Fig. 3.7 An example structure of the supernet model on the server

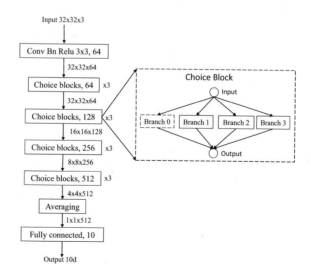

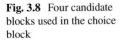

Fig. 3.8 Four candidate blocks used in the choice block

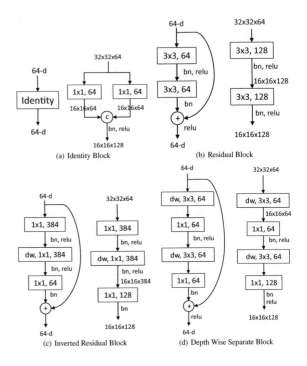

Identity block indicates its input and output are linked directly. Its basic functionality can be seen as a 'layer removal' operation, as shown in the left panel of Fig. 3.8(a). In addition, the right panel of Fig. 3.8(a) represents a reduction block, which contains two branches of convolutional layers with a stride of 2 at first and then concatenates these two outputs.

Residual block is composed of two sequentially connected convolutional blocks as presented in Fig. 3.8(b). The structure of residual block is the same as the one used in ResNet [15], except that the reduction block does not apply to the shortcut connection.

Inverted residual block was first proposed in MobilenetV2 [16] containing three sequentially connected convolutional layers: a 1×1 expanding point-wise convolutional layer, a 3×3 depthwise convolutional layer [17, 18] and a 1×1 spatially filter point-wise convolutional layers.

The depthwise separable block consists of two 3×3 depthwise convolutional layers, each of which is followed by a 1x1 point-wise convolutional layer as shown in Fig. 3.8(c). The advantage of using this block is that it consumes less computational time with the cost of a small performance deterioration.

The network encoding method is very straight-forward, and the entire binary chromosome has a total length of $2 \times 12 = 24$ bits, since the supernet contains 12 choice blocks. Every two bits within the chromosome encodes one specific branch in the choice block. For instance, [0, 0] represents branch 0, which is the identity block, [0, 1] represents branch 1, which is the residual block.

3.3.2 Network Sampling and Client Sampling

In order to develop an RT-FedEvoNAS algorithm, a double sampling technique, including network sampling and client sampling, is adopted in the evolutionary federated learning system. Network sampling randomly sub-samples one branch of each choice block from the global supernet model for each individual in the population according to the generated choice key. Client sampling means all the connected clients are randomly and evenly allocated into groups, the number of which is equal to the population size. And then, each sub-sampled model will be downloaded to the clients belong to a random group without replacement. The number of clients L for training one sub-model of an individual determined by the ratio between the number of individuals N and the number of participants m, i.e., $L = \lfloor m/N \rfloor$. To ensure $L \geq 1$, the number of clients is set to be equal to or larger than that of the population size.

Two objectives will be optimized in the experiments: one is the validation error, the other is the number of parameters of the sub-sampled model. And it is natural to add one more objective in evolutionary multi-objective optimization [19]. Consequently, three objectives, model accuracy, FLOPs and the number of model parameters are included in our comparative studies. Population initialization integrated with network sampling and client sampling can be summarized into the following four steps:

- Initialize the global supernet model on the server. Generate N parent solutions represented by N choice keys, each of which indicates a sub-model sampled from the supernet. Perform client sampling to evenly allocate m connected clients into N groups without replacement, thus, each group includes L clients.
- Sub-sample N sub-models based on the choice keys and download them to the clients of N groups. And then each client can only receive one sub-model for local training. Once the training is completed, every client uploads updated sub-model to the server for aggregation to upgrade the global supernet model.
- Generate N offspring choice keys by crossover and mutation. Similarly, generate N sub-models according to the choice keys and download them to group clients. And then, each client uploads the trained local sub-models to the server for the global model update.
- Finally, download the global supernet together with the choice keys of all parents and offspring to all clients to evaluate their fitness values such as validation errors, FLOPs, and the number of sub-model parameters. After that, all clients upload computed fitness values to the server for fitness evaluations.

Once the objective values of all parents and offspring individuals are calculated, environmental selection can be carried out to generate parent individuals of the next generation based on the elitist non-dominated sorting and crowding distance proposed in NSGA-II. One can also choose any other environmental selection algorithms if required.

Similar optimization steps are adopted in the following generations, with the only difference being that the global supernet model is updated once at each generation after local sub-models of all offspring individuals are trained on the sam-

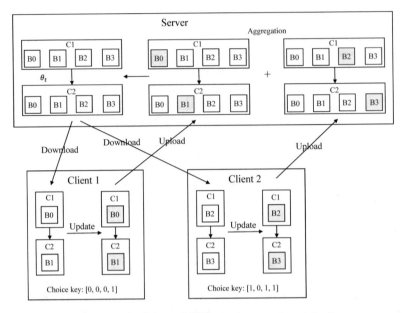

Fig. 3.9 An illustrative example of the model filling and aggregation method

pled clients. Besides, the conventional model aggregation method is not suited for RT-FedEvoNAS, since sub-models uploaded from different group clients cannot be directly aggregated due to their structure differences. Therefore, it is necessary to develop a model filling and aggregation algorithm to solve this issue. As an example, Fig. 3.9 plots a global supernet that has two choice blocks C1 and C2, and the choice keys for two individuals are [0, 0, 0, 1] and [1, 0, 1, 1], respectively.

The sampled sub-models are B0 and B1 on Client 1, and B2 and B3 on Client 2. Both clients operate local training with their local data and upload updated sub-models (denoted by shaded squares) to the server. The server can reconstruct two global models with the same structure by filling the received sub-models with not updated branches (white squares). Thus, the reconstructed global models can be easily aggregated using the FedAvg algorithm. It should be emphasized that the proposed filling and aggregation strategy is performed on the server and it does not incur extra communication costs. The pseudo-code for model aggregation is presented in Algorithm 4.

3.3.3 Overall Framework

NSGA-II is selected as the multi-objective evolutionary algorithm to optimize both sub-model complexity and the validation performance of the RT-FedEvoNAS framework. The overall framework is illustrated in Fig. 3.10 and the pseudo-code is shown in Algorithm 5.

Algorithm 4 Filling and Aggregation Algorithm, I is the total number of hidden layers of the global model, i is the hidden layer index of the learning model, B is the total branches in one choice block, b is the branch block index.

1: $\theta(t-1)$ is the parameters of the global model in the last communication round, $\theta_b^i(t)$ is the parameters of the global model for the b-th branch in the i-th hidden layer.
2: Receive client model parameters θ_k and choice key C_k, where θ_k^i is the model parameters and C_k^i is the choice of i-th hidden layer.
3: **Server Aggregation:**
4: Let $\theta(t) \leftarrow 0$
5: **for** each $i \in I$ **do**
6:     **for** each $k \in m$ **do**
7:         **if** θ_k^i is not in choice blocks **then**
8:             $\theta^i(t) \leftarrow \sum_{k=1}^{m} \frac{n_k}{n} \theta_k^i$
9:         **else**
10:             **for** each branch $b \in B$ **do**
11:                 **if thenC** $C_k^i == b$
12:                     $\theta_b^i(t) \leftarrow \theta_b^i(t) + \frac{n_k}{n} \theta_k^i$
13:                 **else**
14:                     $\theta_b^i(t) \leftarrow \theta_b^i(t) + \frac{n_k}{n} \theta_b^i(t-1)$
15:                 **end if**
16:             **end for**
17:         **end if**
18:     **end for**
19: **end for**
20: **Return** $\theta(t)$

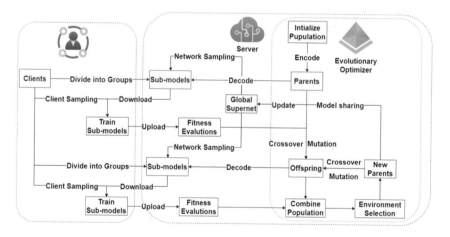

Fig. 3.10 The overall framework of RT-FedEvoNAS

Algorithm 5 RT-FedEvoNAS by NSGA-II

1: **Server:**
2: Initialize $\theta(t = 0)$
3: Randomly sub sample parent choice keys $C_P(t)$ with a population size N
4: **for** each communication round $t = 1, 2, \ldots$ **do**
5:      Convert all $C_P(t)$ choice keys into binary codes $Cb_P(t)$
6:      Generate $Cb_Q(t)$ with the size of N by binary genetic operators
7:      Convert $Cb_Q(t)$ into offspring choice keys $C_Q(t)$
8:      $C_R(t) \leftarrow C_P(t) + C_Q(t)$
9:      Select $m = C \times K$ clients, $C \in (0, 1)$
10:     **if** $t \leq 1$ **then**
11:         Generate sub models $\theta^p \in \theta(t)$, $p \in C_P(t)$
12:         Randomly sub sample m clients to N groups
13:         **for** $p \in C_P(t), g \in N$ $(|C_P(t)| = N)$ **do**
14:             Download θ^P to all clients in group g
15:         **end for**
16:         **for** each client $k \in m$ **do**
17:             Wait **Client k Update** for synchronization
18:             $\theta(t) \leftarrow Do\ server\ aggregation\ in$ **Algorithm 4**
19:         **end for**
20:     **end if**
21:     Generate sub models $\theta^q \in \theta(t)$, $q \in C_Q(t)$
22:     Randomly sub sample m clients to N groups
23:     **for** $q \in C_Q(t), g \in N$ $(|C_Q(t)| = N)$ **do**
24:         **if** $t > 1$ **then**
25:             Download the choice key q to all clients in group g
26:         **else**
27:             Download θ^q and the choice key q to all clients in group g
28:         **end if**
29:     **end for**
30:     **for** each client $k \in m$ **do**
31:         Wait **Client k Update** for synchronization
32:         $\theta(t) \leftarrow Do\ server\ aggregation\ in$ **Algorithm 4**
33:     **end for**
34:     Calculate FLOPs of all sub models in $C_R(t)$
35:     **for** each client $k \in m$ **do**
36:         Download the global supernet $\theta(t)$ and choice keys $C_R(t)$ to client k
37:         Calculate the validation errors for all sub models in $C_R(t)$
38:         Upload them to the server
39:     **end for**
40:     Do weighted averaging on validation errors of all uploads based on the local data size to achieve the final validation
        errors of all sub models in $C_R(t)$
41:     Do fast non dominated sorting
42:     Do crowding distance sorting
43:     Generate new parent choice keys $C_P(t)$ from $C_R(t)$
44:     $t \leftarrow t + 1$
45: **end for**
46:
47: **Client k Update:**
48: **if** Receive one choice key q **then**
49:     Sub sample $\theta(t)$ based on the choice key to generate θ^k
50: **else**
51:     $\theta^k \leftarrow \theta^p or \theta^q$
52: **end if**
53: **for** each iteration from 1 to E **do**
54:     **for** batch $b \in B$ **do**
55:         $\theta^k = \theta^k - \eta \nabla L_k(\theta^k, b)$
56:     **end for**
57: **end for**
58: return θ^k to the server

Both the global supernet model and the choice keys $C_R(t)$ of the entire population should be downloaded to all connected clients for fitness evaluations. As a result, only the choice keys are required to be downloaded at the next round (refer to line 25 in Algorithm 5), since the entire supernet has already been downloaded at the last round. And then, each group client uploads the sub-model to the server after local training, significantly reducing the local computational overhead and uploading costs. Besides, the parent sub-models are trained only at the first generation, and only offspring sub-models need to be trained in the subsequent evolutionary generation. On the other hand, fitness computations are performed on the entire $2N$ individuals at each generation, since training the offspring sub-models will also affect the parameters of the parent sub-models due to applied network sharing technique.

A set of Pareto optimal sub-models can be found at each evolutionary generation. Thus, the server requires to articulate preferences to select one or more solutions from the parent population and distributes them to clients for the real-time usage. In general, the the Pareto optimal solutions with the highest validation accuracy and the knee point [20–22] are preferred unless there are other strong user-specified preferences.

3.3.4 Empirical Studies

The empirical studies here are used to verify that the proposed RT-FedEvoNAS is able to accomplish the real-time federated NAS tasks. Three datasets, CIFAR10 [23], CIFAR100 and SVHN are used in our experiments. To simulate IID data distributions, all training images are evenly and randomly distributed to each client without overlap. For experiments on non-IID data, each client has images with five classes of objects for the CIFAR10 and SVHN datasets, and 50 classes of objects for the CIFAR100 dataset.

The global supernet contains 12 choice blocks, each of which has 4 branches and the number of channels for all choice blocks are [64, 64, 64, 128, 128,128, 256, 256, 256, 512, 512, 512], respectively. For brevity, both trainable parameters and moving statistics in the batch normalization layers are disabled. And the sub-models optimized by the proposed algorithm is compared with five predefined baseline models over the entire evolutionary generations. These baseline models are ResNet18, MobileNet, MobileNetv2, and PnasNet-1 and PnasNet-2 searched by PNASNET [24]. Other experimental settings are listed in Table 3.3.

3.3.4.1 Results on Different Numbers of Clients

In this set of experiments, we examine the performance of the obtained sub-models with four different numbers of clients (10, 20, 50 and 100) and optimize their validation errors and the model FLOPs. The obtained Pareto optimal solutions on Non-IID data over every 50 generations are shown in Fig. 3.11.

Table 3.3 Settings of the experiments

Setting	Value
Generations	300
Population	10
Crossover probability	0.9
Mutation probability	0.1
Bit length	2
Total clients	10, 20, 50, 100
Local epochs	1
Batch Size	50
Initial learning rate	0.1
Momentum	0.5
Learning decay	0.995

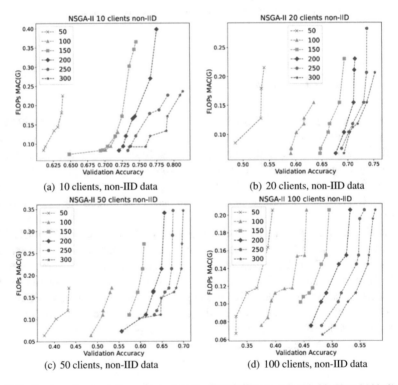

Fig. 3.11 Pareto optimal solutions obtained on the CIFAR10 dataset for 10, 20, 50 and 100 clients. The integer numbers in the legend in each subplot represent the number of communication rounds (generations) during the real-time federated evolutionary optimization

Table 3.4 Final test performances of the evolved two Pareto solutions on CIFAR10

Model	Clients	IID	Test accuracy (%)	FLOPs (MAC)
High	10	No	**76.51**	0.2356G
Knee	10	No	76.11	**0.0920G**
High	20	No	**78.16**	0.2060G
Knee	20	No	76.28	**0.1542G**
High	50	No	**73.66**	0.3465G
Knee	50	No	71.21	**0.1702G**
High	100	No	**63.98**	0.2055G
Knee	100	No	62.19	**0.1443G**

It is obvious to find that the smaller the number of clients, the better the classification accuracy, and the population converges well across generations. To take a closer look at the final results of the obtained sub-models on the CIFAR10 dataset, both test accuracy and FLOPs of the sub-models with the highest validation accuracy and those near the knee point are listed in Table 3.4. It should be emphasized that the test accuracy of the found sub-models are directly calculated on the test data without performing any re-training operations.

From these results, it is clear to see that solutions with higher model FLOPs always have better final test accuracy. Furthermore, the final test accuracy of the sub-models trends to become lower as the number of clients increases. For example, the accuracy is 76.51% accuracy for 10 clients and 63.98% for 100 clients. This implies that it becomes harder to find an optimal sub-model as the amount of data on each client becomes less.

3.3.4.2 Real-Time Performance

Apart from the final test model performance and the convergence of evolutionary optimization, the real-time performance of the obtained sub-models at each generation is also investigated in our experiments. And an extra objective, the number of model parameters, is added to the previous bi-optimization problem to make it a three-objective optimization problem. Similarly, two Pareto optimal sub-models are selected to compare objective values over communication rounds with the baseline models. The number of participants are set to be 20, and the results are shown in Fig. 3.12.

For the CIFAR10 dataset, two solutions (High and Knee) converge slower than the baseline models at the beginning of the steps. Nevertheless, two solutions perform better than PnasNet-2 and MobileNet at the middle of rounds and the High solution outperforms all baseline models in the end with a validation accuracy of 74.37%. Besides, PnasNet-1 and PnasNet-2 have worse performance than other models with a validation accuracy of 69.74% and 62.50%, respectively.

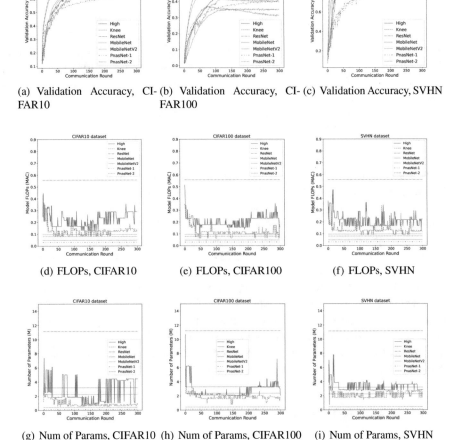

(a) Validation Accuracy, CI- (b) Validation Accuracy, CI- (c) Validation Accuracy, SVHN
FAR10   FAR100

(d) FLOPs, CIFAR10   (e) FLOPs, CIFAR100   (f) FLOPs, SVHN

(g) Num of Params, CIFAR10 (h) Num of Params, CIFAR100 (i) Num of Params, SVHN

Fig. 3.12 Real-time performances of two Pareto optimal solutions and five baseline models over
the rounds (generations)

The curves of validation accuracy for CIFAR100 dataset look similar to those
for CIFAR10 dataset, and both High and Knee sub-models outperform the baseline
models with a validation accuracy around 53% at the middle stage (Fig. 3.12(b)). It
is easy to find that the validation accuracy of two solutions are much more stable
than five baseline models over the communication rounds, and ResNet, MobileNet,
MobileNetV2 and PnasNet-2 have a dramatic performance drop at a quarter of the
total training rounds. This phenomenon indicates that these baseline models are not
well suited for real-time federated learning tasks on complicated datasets such as
CIFAR100, and the advantage of our proposed real-time federated NAS algorithm
is apparent.

The results on the SVHN dataset are presented in Fig. 3.12(c). Except for PnasNet-1, all comparative models and two sub-models found by RT-FedEvoNAS converge very fast and exhibit good final performances. The validation accuracy of PnasNet-1 is much lower than that of the other models due to its low model complexity.

Next, the complexity of the derived sub-models in terms of FLOPs and the number of parameters will be compared with that of five baseline models. The models FLOPs on three datasets are shown in Fig. 3.12(d)(e)(f), respectively. The average FLOPs of the High and the Knee sub-models are 0.26G and 0.13G, while the average FLOPs of five baseline models are about 0.56G, 0.05G, 0.10G, 0.03G and 0.08G, respectively. Furthermore, the number of model parameters are shown in Fig. 3.12(g)(h)(i), respectively. It is clear to see that two obtained sub-models have a much fewer number of model parameters than ResNet, although they outperform ResNet on all three datasets. And the real-time model parameters of the High solution fluctuated heavily on CIFAR10 and SVHN, while it does not change much on CIFAR100. This implies that on simpler training datasets such as CIFAR10 and SVHN, smaller models may have similar performance to larger models. In addition, the number of parameters of the Knee solutions is between MobileNetv2 and PnasNet-2 over most of the communication rounds.

Overall, our proposed RT-FedEvoNAS approach can find sub-models with competitive real-time learning performance and a lightweight model size.

3.3.4.3 Comparison with Offline Federated NAS

It is meaningful to compare our proposed real-time federated NAS algorithm with the aforementioned offline evolutionary approach [9]. However, it is hard to make a fair comparison of these two methods even if the same search space is adopted. This is because the offline approach trains the individual's sub-model on all connected clients and the number of communication rounds used for fitness evaluations has to be manually set. And a larger number of communication rounds make each sampled neural network well trained, but consuming much more local computational and communication resources. In addition, the solutions searched by the offline NAS method often need to be trained again, which also incurs extra computation resources and communication costs.

For the offline method, the number of communication rounds for training one individual sub-model is set to 10, and the total number of generations is set to be 20. Although the total number of generations of the offline method is much smaller than that of RT-FedEvoNAS, the actual number of communication rounds used in the offline evolutionary search is $10 \times 10 \times 20 = 2000$ (the population size is 10, 10 communication rounds and 20 generations), which is much larger than 300 rounds for RT-FedEvoNAS.

The Pareto optimal solutions found over 50 generations by RT-FedEvoNAS and those found by the offline method in the last generation are plotted in Fig. 3.13. And the solutions found by the real-time method at the 100th generation dominates the final solutions obtained by the offline method at the 2000th generation. This

Fig. 3.13 Pareto optimal
solutions where the integer
numbers in the legend denote
the number of generations

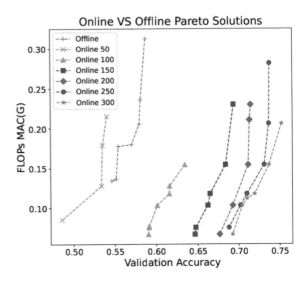

Table 3.5 Time consumption and number of downloaded and uploaded model parameters for RT-FedEvoNAS and offline evolutionary federated NAS

Methods	RT-FedEvoNAS	Offline
Avg Time consumption per generation	2 min	162 min
Avg Client Downloads per generation	31.32 M	715.98 M
Avg Client uploads per generation	6.19 M	715.98 M

indicates that the solutions found by the offline method is over-fitting and requires further training, while the solutions found by our proposed real-time method has already been well trained.

The average time consumption and communication costs per evolutionary generation are listed in Table 3.5. From the table, we see that the average time consumption per generation of the offline approach is around 81 times larger than that of the proposed real-time method. At the same time, the average numbers of model parameters downloaded and uploaded between the server and each client are 31.32 M and 6.19 M for RT-FedEvoNAS, respectively, which are 23 times and 116 times less than those for the offline method. Therefore, the proposed RT-FedEvoNAS consumes much less computation and communication resources compared to the offline federated NAS method.

3.4 Summary

In this chapter, we introduce two evolutionary federated learning algorithms. One is an offline multi-objective evolutionary federated neural architecture search algorithm, and the other is a method for real-time federated evolutionary neural architecture search. The first method directly optimizes the neural structure used in federated learning to simultaneously reduce the communication costs and enhance the global model performance. It is inherently an offline approach in that the structure optimization and model training are performed separately and that each client needs to train multiple local models in one round. The local computational and communication overhead is heavily dependent on the population size.

Therefore, a real-time federated evolutionary neural architecture search method is proposed to overcome these issues. It adopts network sampling to reduce both communication costs and local computational overhead. And the introduced model sharing technique makes it possible to avoid dramatic performance drop during the optimization. In addition, client sampling is used to ensure one evolutionary generation is accomplished within one communication round, thus without requiring each clients to train multiple models in each round.

However, the proposed RT-FedEvoNAS suffers from certain limitations. The global model is restricted to a particular machine learning model types (e.g., neural networks) with a specific network structure (such as the multi-branch supernet structure). Moreover, client sampling may bring in fitness evaluation bias, which becomes more severe on non-IID client data. In the next chapter, we will present federated learning algorithms that use non-parametric machine learning models for vertical data partitions.

References

1. Jin, Y., Sendhoff, B.: Pareto-based multiobjective machine learning: an overview and case studies. IEEE Trans. Syst. Man Cybern. Part C (Applications and Reviews) **38**(3), 397–415 (2008). https://doi.org/10.1109/TSMCC.2008.919172
2. Mocanu, D.C., Mocanu, E., Stone, P., Nguyen, P.H., Gibescu, M., Liotta, A.: Scalable training of artificial neural networks with adaptive sparse connectivity inspired by network science. Nat. Commun. **9**(1), 1–12 (2018)
3. Erdős, P., Rényi, A., et al.: On the evolution of random graphs. Publ. Math. Inst. Hung. Acad. Sci **5**(1), 17–60 (1960)
4. Deb, K.: Multi-objective optimization. In: Search Methodologies, pp. 403–449. Springer, Berlin (2014)
5. McMahan, B., Moore, E., Ramage, D., Hampson, S., Arcas, B.A.Y.: Communication-efficient learning of deep networks from decentralized data. In: Singh, A., Zhu, J. (eds.) Proceedings of the 20th International Conference on Artificial Intelligence and Statistics, Proceedings of Machine Learning Research, vol. 54, pp. 1273–1282. PMLR (2017). https://proceedings.mlr.press/v54/mcmahan17a.html
6. Deb, K., Pratap, A., Agarwal, S., Meyarivan, T.: A fast and elitist multiobjective genetic algorithm: NSGA-II. IEEE Trans. Evol. Comput. **6**(2), 182–197 (2002). https://doi.org/10.1109/4235.996017

7. Deb, K., Agrawal, R.B., et al.: Simulated binary crossover for continuous search space. Complex Syst. **9**(2), 115–148 (1995)
8. Lecun, Y., Bottou, L., Bengio, Y., Haffner, P.: Gradient-based learning applied to document recognition. Proc. IEEE **86**(11), 2278–2324 (1998). https://doi.org/10.1109/5.726791
9. Zhu, H., Jin, Y.: Multi-objective evolutionary federated learning. IEEE Trans. Neural Netw. Learn. Syst. **31**(4), 1310–1322 (2020). https://doi.org/10.1109/TNNLS.2019.2919699
10. Suganuma, M., Shirakawa, S., Nagao, T.: A genetic programming approach to designing convolutional neural network architectures. In: Proceedings of the Genetic and Evolutionary Computation Conference, GECCO '17, pp. 497–504. Association for Computing Machinery, New York, NY, USA (2017). https://doi.org/10.1145/3071178.3071229, https://doi.org/10.1145/3071178.3071229
11. Angeline, P., Saunders, G., Pollack, J.: An evolutionary algorithm that constructs recurrent neural networks. IEEE Trans. Neural Netw. **5**(1), 54–65 (1994). https://doi.org/10.1109/72.265960
12. Lu, Z., Whalen, I., Boddeti, V., Dhebar, Y., Deb, K., Goodman, E., Banzhaf, W.: NSGA-net: neural architecture search using multi-objective genetic algorithm. In: Proceedings of the Genetic and Evolutionary Computation Conference, GECCO '19, pp. 419–427. Association for Computing Machinery, New York, NY, USA (2019). https://doi.org/10.1145/3321707.3321729, https://doi.org/10.1145/3321707.3321729
13. Jin, Y., Wang, H., Chugh, T., Guo, D., Miettinen, K.: Data-driven evolutionary optimization: an overview and case studies. IEEE Trans. Evol. Comput. **23**(3), 442–458 (2019). https://doi.org/10.1109/TEVC.2018.2869001
14. Ioffe, S., Szegedy, C.: Batch normalization: accelerating deep network training by reducing internal covariate shift. In: Bach, F., Blei, D. (eds.) Proceedings of the 32nd International Conference on Machine Learning, Proceedings of Machine Learning Research, vol. 37, pp. 448–456. PMLR, Lille, France (2015). https://proceedings.mlr.press/v37/ioffe15.html
15. He, K., Zhang, X., Ren, S., Sun, J.: Deep residual learning for image recognition. In: Proceedings of the IEEE Conference on Computer Vision and Pattern Recognition, pp. 770–778 (2016)
16. Sandler, M., Howard, A., Zhu, M., Zhmoginov, A., Chen, L.C.: Mobilenetv2: Inverted residuals and linear bottlenecks. In: Proceedings of the IEEE Conference on Computer Vision and Pattern Recognition, pp. 4510–4520 (2018)
17. Chollet, F.: Xception: Deep learning with depthwise separable convolutions. In: Proceedings of the IEEE Conference on Computer Vision and Pattern Recognition, pp. 1251–1258 (2017)
18. Howard, A.G., Zhu, M., Chen, B., Kalenichenko, D., Wang, W., Weyand, T., Andreetto, M., Adam, H.: Mobilenets: Efficient convolutional neural networks for mobile vision applications (2017). arXiv:1704.04861
19. Coello Coello, C.A., González Brambila, S., Figueroa Gamboa, J., Castillo Tapia, M.G., Hernández Gómez, R.: Evolutionary multiobjective optimization: open research areas and some challenges lying ahead. Complex Intell. Syst. **6**(2), 221–236 (2020)
20. Yu, G., Jin, Y., Olhofer, M.: Benchmark problems and performance indicators for search of knee points in multiobjective optimization. IEEE Trans. Cybern. **50**(8), 3531–3544 (2020). https://doi.org/10.1109/TCYB.2019.2894664
21. Yu, G., Jin, Y., Olhofer, M.: A method for a posteriori identification of knee points based on solution density. In: 2018 IEEE Congress on Evolutionary Computation (CEC), pp. 1–8 (2018). https://doi.org/10.1109/CEC.2018.8477885
22. Zhang, X., Tian, Y., Jin, Y.: A knee point-driven evolutionary algorithm for many-objective optimization. IEEE Trans. Evol. Comput. **19**(6), 761–776 (2015). https://doi.org/10.1109/TEVC.2014.2378512
23. Krizhevsky, A., Hinton, G.: Learning multiple layers of features from tiny images. Master's thesis, Department of Computer Science, University of Toronto (2009)
24. Liu, C., Zoph, B., Neumann, M., Shlens, J., Hua, W., Li, L.J., Fei-Fei, L., Yuille, A., Huang, J., Murphy, K.: Progressive neural architecture search. In: Proceedings of the European Conference on Computer Vision (ECCV), pp. 19–34 (2018)

Chapter 4
Secure Federated Learning

Abstract Although federated learning does not require any participant to share its private data to the central server, local data information still has a potential risk of being revealed taking advantage of the uploaded model parameters or gradients from each client, if the server is *honest-but-curious*. To address this issue, homomorphic encryption is one of the main stream privacy preservation technologies used in federated learning due to its extraordinary model protection ability. However, most encryption-based federated learning systems are computationally intensive and consume a large number of communication resources. Moreover, a trusted third party is always needed to generate key pairs for both encryption and decryption operations, not only increasing the complexity of the system topology but also causing additional security threats. Therefore, in this chapter, we introduce two secure federated learning frameworks. The first one is for horizontal federated learning, where the global key pairs are jointly generated between the server and connected clients without the help of a trusted third party. In addition, model quantization and approximated model aggregation techniques are adopted to significantly improve the encryption efficiency during the training period. The second framework is tailored for vertical federated learning, where labels are distributed on different local devices. To achieve secure node splitting and construction as well as label aggregation of a XGBoost tree, both homomorphic encryption and differential privacy are adopted.

4.1 Threats to Federated Learning

Federated learning (FL) [1] is an emerging distributed machine learning framework designed for the purpose of privacy preservation. It allows multiple parties to collaboratively train a shared global model, while keeping all the private data stored locally on the device. Consequently, FL has already been applied in many real-world scenarios like mobile platforms [2–5], healthcare [6–16] and other industrial sectors. However, some recent studies [17–22] have indicated that a individual client's private data still risks being revealed in FL, since the uploaded model parameters or gradients contain knowledge about their input training data.

The first adversarial method on collaborative deep learning was proposed in [23], where a malicious client secretly builds a local generative adversarial net (GAN) [24] to recover digit images of a specific label that it does not have. However, this approach requires to change the architecture of shared model classifier, making it unrealistic in practical scenarios. Based on aforementioned work, Hardy et al. introduce a novel framework called Multi-Discriminator GAN (MD-GAN) [25] to reduce half of local computational overheads by putting the generator only in the server. But this scheme requires connected clients to perform peer-to-peer discriminator exchanges, which significantly increases communication costs. At the same time, Wang et al. developed a mGAN-AI system [26] to enable an adversarial server to recover user-level private information of a specific client. More recently, Zhang et al. [27] adopt a GAN to generate attack data for enhancing the membership inference attack. Consequently, it can be deduced whether a data sample exists or not within the training data of a client.

Apart from GAN based attacks, inverting gradients is a more direct adversarial technique to extract data privacy through uploaded model gradients. Geiping et al. [28] successfully recover a high-resolution image from the model gradients by optimizing a constructed magnitude-invariant loss. Experimental results indicate that user's privacy in FL cannot be fully protected even averaging the local gradients of 100 images. Similar work on image recovery is also proposed in [29] by shortening the Euclidean distance between the gradients of dummy data and the real training data. After that, Lim and Chan [30] present a more detailed study and analysis on the impact of gradient inversion issues in FL and give a novel perspective view on the reliability of DNN feature learning. Although these gradient inversion attacks are performed with some strong assumptions [31], local data privacy is under threat to a certain degree.

Therefore, additional privacy protection technologies like differential privacy (DP) [32], secure multi-party computations (SMC) [33], homomorphic encryption (HE) [34] are often required to further enhance the security level of FL systems.

4.2 Distributed Encryption for Horizontal Federated Learning

Homomorphic encryption (HE) [34, 35] is a very powerful tool for protecting communicated model parameters or gradients in horizontal federated learning (HFL). It naturally supports ciphertext computations, thus, the encrypted model gradients from the connected clients can be aggregated directly on the server without first decrypting them. The aggregated global model is still in a ciphertext form, which, when decrypted, outputs the same model obtained on the unencrypted local models.

Some work have already been done to apply HE in HFL systems. Phong et al. [19] first adopt learning with error (LWE) based and Paillier additive HE [35] to protect the shared model gradients in distributed deep learning and they also theoretically prove that the model gradients are indeed proportional to the input training data. Later, Truex et al. combine SMC and Paillier encryption to produce a more practical threshold based HE FL system [36], in which the key pairs are distributively generated among participants. Similar methods are also used in [37, 38], which, however, dramatically increases the communication costs consumed in FL, since the encrypted data always have a large bit length for security concerns. Recently, Zhang et al. conducted a novel batchcrypt scheme [39] that not only decreases the communication costs but also accelerates the local encryption speed. Instead of encrypting the full precision elements of model gradients one by one, batchcrypt encodes a batch of quantized gradients into a longer vector and encrypts the vector once as a whole.

However, the above mentioned HE based FL systems require an extra trusted third party (TTP) [40] to generate key pairs, which deteriorates the topological structure of HFL. Moreover, using only one key pair in HFL is not well suited for real-world FL systems, since the central server can easily decrypt the uploaded ciphertexts if it receives the secret key from one of adversarial clients. Although the introduced threshold based approach [36] can deal with this threat, it consumes much more communication resources due to distributed decryption operations.

Therefore, it is necessary to develop an efficient threshold based HE system for HFL, where the global key pairs are generated on distributed devices without the help of TTP. In this case, no party will hold the global secret key and the security level of the whole FL system can be significantly enhanced.

4.2.1 Distributed Data Encryption

Distributed data encryption and decryption algorithms are originally designed to deal with distributed data storage systems. Postma et al. [41] first propose a threshold based distributed Rivest–Shamir–Adleman (RSA) [42] cryptosystem, in which both the public key and private key are decomposed into n key fragments. These key fragments are distributed on a number of nodes (each node may possess several fragments) to perform the encryption function upon partially encrypted data shards. However, how to distributedly generate key pairs are not explicitly described in that work. Agrawal et al. also utilize the mechanism of threshold based cryptography to build a distributed symmetric-key encryption framework [43], which works in the following way. If a client wants to encrypt a message, it will contact t out of n servers for the secret key shards (servers do not contact with each other). Then, the client can combine the received key shards to generate the key for message encryption. Unfortunately, the system topology of proposed framework is not applicable to HFL.

Instead of the aforementioned approaches, using only one key pair for distributed data encryption and decryption is a more general method. However, this type of methods is not robust to adversarial attacks and always requires the (central) parameter server to be honest. As shown in [44] that all the training data and model parameters are stored on the parameter server, in which the global key pairs are created. Both data and the model are encrypted by the generated public key and sent to each connected worker node. After that, every worker node can update the received ciphertexts due to the mechanism of homomorphism without achieving any knowledge of the plaintext data. A similar idea is also adopted in [45], however in this case, the input data are stored on distributed devices. Therefore, the proposed system needs a crypto-service provider (actually a TTP) for both key pair generations and collaborative computing. Moreover, Aono et al. [46] introduced a distributed logistic regression system [47] by applying several HE algorithms. In that work, a randomly chosen client is responsible for key pair generations and the created public key should be distributed to any other clients. Although this scheme does not use a TTP for key generation and distribution, the selected client must be benign and honest.

To the best of our knowledge, distributed key generation (DKG) [48] is the most suitable technology for secure key generations in HFL. Based on the principle of Shamir Secret Sharing (SSS) [49] and Diffie–Hellman (DH) definition, the global key pairs can be collaboratively created among multiple participants without the help of TTP. In addition, this scheme is robust to adversarial participants, who intentionally upload incorrect messages to break down the key pair generation, by adopting two verification processes.

4.2.2   Federated Encryption and Decryption

The most challenging part for performing practical federated encryption and decryption system is generating the global key pairs without the assistance of a TTP. In addition, the proposed system should also be robust to adversarial clients who may upload wrong messages to impede successful key generation. Therefore, a variant of DKG called federated key generation (FKG) is a central component for this system, allowing the server and clients to jointly create the desired global key pairs. And discrete logarithms based ElGamal HE cryptosystem [50] can be applied together with FKG, since it perfectly matches the scope of FKG. However, ElGamal is inherently a multiplicative HE scheme. Consequently, an extra transformation is required to convert ElGamal into an additive HE, making it well suited for model aggregations in HFL. Then, a simple fixed point encoding method will be introduced, followed by two different recovery approaches to retrieve the desired plaintext messages after decryption.

4.2.2.1   Threat Model

Before introducing the proposed FKG framework, the threat model is defined at first. The potential attack sources in HFL can come from the server, participating clients and outsiders. And existing FL systems often ignore outside attackers and assume that the server is *honest-but-curious* [51, 52]. To be more specific, at least t-out-of-n clients are considered to be honest in the system. Thus, the remaining $n - t$ clients might be malicious and try to deduce data information from other honest clients.

Therefore, our main goal is to prevent client data information from being revealed without considering model skewing attacks in FL.

4.2.2.2 Federated Key Generation

The proposed federated key generation (FKG) originates from distributed key generation (DKG), and the main steps are introduced in Algorithm 1.

Algorithm 1 Federated Key Generation. $\mathbb{G}$ is a cyclic group, p and q are large prime numbers, g is a generator, $y \in \mathbb{G}$, N is the number of total clients, C is the fraction of clients participating the current round, i is the client index, and T is the threshold value

---

1: *Server distributes public parameters* $< p, q, g, y >$
2: **for** Each FL round $t = 1, 2, \ldots$ **do**
3:     $n = C * N$
4:     Select threshold value $T > n/2$
5:     Client $i \in \{1, \cdots, n\}$ perform **Pedersen VSS**
6:     Collect the number of complaints cpt_i for client i
7:     **for** Client $i \in \{1, \cdots, n\}$ **do**
8:         **if** $cpt_i > T$ **then**
9:             Mark client i as disqualified
10:         **else**
11:             Client i uploads $f_i(j)$ and
12:             **if** Eq. (4.2) is satisfied **then**
13:                 Mark client i as qualified (QUAL)
14:             **else**
15:                 Mark client i as disqualified
16:             **end if**
17:             Mark client i as QUAL   ▷ $T \leq |\text{QUAL}| \leq n$
18:         **end if**
19:     **end for**
20:     Client $i \in$ QUAL perform **Feldman VSS**
21:     Collect complained client index in O,
22:     **for** Each client $i \in O$ **do**   ▷ $|O| < T$
23:         Set $counter = 0$
24:         **for** Each client $j \in$ QUAL but $j \notin m$ **do**
25:             Client j uploads $f_j(i)$ and $f'_j(i)$
26:             **if** Eq. (4.2) is satisfied **then**
27:                 $counter = counter + 1$
28:             **end if**
29:             **if** $counter \geq T$ **then**
30:                 Break
31:             **end if**
32:         **end for**
33:         Retrieve $f_i(z)$ and A_{i0} ▷ $A_{i0} = g^{a_{i0}} \pmod{p}$
34:     **end for**
35:     Generate global public key $h = \prod_{i \in QUAL} A_{i0} = g^x$
36: **end for**

Four public key parameters p, q, g and y are generated on the server before distributing them to each client, where q is the prime order of cyclic group $\mathbb{G}$, p is a large prime number satisfying $p - 1 = rq$, r is a positive integer, g and y are two different random elements in $\mathbb{G}$.

In order to ensure the generated key pairs are correct, Pedersen verifiable secret sharing (VSS) and Feldman verifiable secret sharing (VSS) are used in FKG. Pedersen VSS requires each client i to create two random polynomials $f_i(z)$ and $f_i'(z)$ over $\mathbb{Z}_q^*$ as shown in Eq. (4.1).

$$
\begin{aligned}
f_i(z) &= a_{i0} + a_{i1}z + \ldots + a_{iT-1}z^{T-1} \quad (\text{mod } q), \\
f_i'(z) &= b_{i0} + b_{i1}z + \ldots + b_{iT-1}z^{T-1} \quad (\text{mod } q),
\end{aligned}
\tag{4.1}
$$

where $a_{i0} = f_i(0)$ is the private key shard used to distributedly generate the global private key and T is the threshold value. Any client i broadcasts $C_{ik} = g^{a_{ik}} y^{b_{ik}}$ (mod p) and sends two shares $s_{ij} = f_i(j)$, $s_{ij}' = f_i'(j)$ to client j. And then every client j verifies the received two shares by the following Eq. (4.2):

$$
g^{s_{ij}} y^{s_{ij}'} = \prod_{k=0}^{T-1} (C_{ik})^{j^k} \quad (\text{mod } p).
\tag{4.2}
$$

It should be emphasized that any adversarial attackers cannot guess a_{ik} and b_{ik} through C_ik or find another pair of s_{ij} and s_{ij}' satisfying Eq. (4.2) due to the hiding and binding properties of Pedersen commitment [53]. It is also impossible to reconstruct any private key shard of honest clients based on the previously described SSS principle.

If the received shares s_{ij} and s_{ij}' cannot satisfy Eq. (4.2), client j will send a complaint of client i to the server. As long as the server receives more than T complaints against client i, this client will be directly classified as disqualified (line 9 in Algorithm 1). In order to avoid malicious complaints, any client i receives less than T complaints is required to upload shares to the server for verification by Eq. (4.2). If any verification fails, client i will be marked as disqualified.

After qualified (QUAL) clients has been verified by Pedersen VSS, Feldman VSS is needed to guarantee all the QUAL clients broadcast correct A_{i0} (line 33 in Algorithm 1) for the public key generation. For Feldman VSS, each client j ($j \in$ QUAL) broadcasts $A_{ik} = g^{a_{ik}}$ (mod p) and verifies Eq. (4.3).

$$
g^{s_{ij}} = \prod_{k=0}^{T-1} (A_{ik})^{j^k} \quad (\text{mod } p)
\tag{4.3}
$$

If s_{ij} of client i does not satisfy Eq. (4.3), client j will send a complaint to the server. And the server will ask T QUAL clients to share their $f_i(j)$ to retrieve the random polynomial $f_i(z)$ of client i using the Lagrange interpolation function [54] in Eq. (4.4).

$$\lambda_j = \prod_{k \neq j} \frac{z - k}{j - k}, k \in T, j \in \text{QUAL}$$

$$f_i(z) = \sum_{j \in \text{QUAL}} \lambda_j f_i(j) \tag{4.4}$$

Finally, the global public key is distributedly generated through A_{i0} by Eq. (4.5), where x is in fact the global private key and h is the public key. Even g, h and p is known to all users, it is very hard to deduce x.

$$h_i = A_{i0} = g^{z_i} \pmod{p}$$

$$h = \prod_{i \in \text{QUAL}} h_i = g^{\sum_{i \in \text{QUAL}} z_i} = g^x \pmod{p} \tag{4.5}$$

4.2.2.3 Additive Discrete Logarithm Based Encryption

After generating the global public key, each client can encrypt the local model parameters before uploading them to the server. In order to be fully compatible with aforementioned FKG, an additive discrete logarithm based encryption scheme is proposed by altering the ElGamal encryption algorithm [55]. The standard ElGamal encryption includes the following steps:

1. Generate three parameters p, q and g, where q is the prime order of a cyclic group $\mathbb{G}$, p is a large prime number satisfying $q|p - 1$, and g is a generator of $\mathbb{G}$.
2. Select a random number $x, x \in \mathbb{Z}_q^*$ as the secret key, and then compute $h = g^x$ (mod p) to be the public key.
3. Randomly select an integer number $r \in \mathbb{Z}_q^*$ to encrypt message $m \in \mathbb{Z}_p^*$, and two ciphertexts are generated as $< c_1 = g^r \pmod{p}, c_2 = mh^r \pmod{p} >$.
4. The plaintext m can be decrypted by the private key x as shown in Eq. (4.6).

$$\frac{c_2}{c_1^x} = \frac{mh^r}{(g^r)^x} = \frac{mg^{xr}}{g^{rx}} \pmod{p} \equiv m \tag{4.6}$$

It is easy to find that the standard ElGamal is a multiplicative homomorphic encryption (HE), since $\text{Enc}(m1) * \text{Enc}(m2) = m_1 m_2 h^{r_1 + r_2} \pmod{p} = \text{Enc}(m1 * m2)$. However, the model aggregation on the server performs addition (averaging) operations, enabling it applicable for additive HE. Therefore, the Cramer transformation [56] is applied to ElGamal by transforming the message m into $m' = g^m$ (mod p). Consequently, the original ElGamal encryption becomes a discrete logarithm based additive HE, as shown in Eq. (4.7).

$$\text{Enc}(m_1) * \text{Enc}(m_2) = g^{m_1} h^{r_1} * g^{m_2} h^{r_2}$$

$$= g^{m_1 + m_2} h^{r_1 + r_2} \pmod{p} \tag{4.7}$$

4.2.2.4   Fixed Point Encoding Method

Model parameters are required to be encoded into integer numbers before encryption, because HE can only be applied to integers. For brevity, a straightforward encoding method is used here as shown in Eq. (4.8), where r is the real number, b is the encoding bits, q is the prime order of cyclic group $\mathbb{G}$, m is the encoded integer number and int_{max} is the maximum positive encoding number.

$$\widehat{m} = \text{round}(r * 2^b), \ |\widehat{m}| \leq int_{max}$$
$$\text{Encode}(r) = \widehat{m} \pmod{q} = m$$
$$\text{Decode}(m) = \begin{cases} m * 2^{-b}, & m \leq int_{max} \\ (m - q) * 2^{-b}, & m > q - int_{max} \end{cases} \tag{4.8}$$

It should be emphasized that if $\widehat{m}$ is negative, the encoded $m = \widehat{m} + q$ should subtract q before decoding. And the bit length of int_{max} is always set to be much larger than the encoding bit b, so that sufficient value space can be reserved for encoding number summations.

4.2.2.5   Brute Force and Log Recovery

As mentioned in Sect. 4.2.2.3, Cramer transformation is adopted to convert the standard ElGamal HE into additive HE, making it match the scope of model aggregation in HFL. And the aggregated global model, when decrypted, requires to recover the desired message m from transformed message $m' = g^m \pmod{p}$. However, solving this type of discrete logarithm hard problem (DLHP) is time consuming, thus, brute force and log recovery methods are proposed to alleviate this issue.

Brute force recovery, like its name implies, traverses different integers m from 0 to $q - 1$ until the equation $m' = g^m \pmod{p}$ is satisfied. Therefore, the theoretical maximum number of trials is q to solve DLHP in the worst case. However, the value m' is actually determined by the encoding bit length b introduced in Sect. 4.2.2.4, and the maximum number of trials becomes approximately 2^b if $\widehat{m}$ is positive. Consequently, there is a trade-off relationship between the encoding precision and the brute force recovery efficiency. A larger bit length will decrease the encoding bias but slow down the brute force recovery speed, and vice versa.

The log recovery method is much faster than aforementioned brute force approach by computing log function $\log_g(g^m) = m$ directly if $g^m < p$. Since g^m has strict up bound threshold, $g_0 = 2$ is replaced as the base g of the Cramer transformation. As a result, a larger encoding bit length b can be used to enhance the encoding precision on the premise of $g^m < p$. By the way, setting a fixed g_0 does not affect the security level of ElGamal encryption, because the transformed message should multiply $h = g^x \pmod{p}$ to generate one of two ciphertexts, where g is the original generator of $\mathbb{G}$ as introduced in Sect. 4.2.2.3.

Both brute force and log recovery are described in Algorithm 2, where q is a prime order of the cyclic group $\mathbb{G}$, p is a large prime number satisfying $p - 1 | q$, g is a random element in $\mathbb{G}$, m is the message to be recovered, and int_{max} is the maximum positive encoded number.

Algorithm 2 Plaintext Recovery

1: **Brute Force Recovery:**
2: $g_0 = g$
3: Given decrypted plaintext $m' = g_0^m \pmod{p}$, $m \leq int_{max}$
4: **for** j from 0 to $q - 1$ **do**
5:    **if** $g_0^j \pmod{p} == m'$ **then**
6:       $m = j$
7:       Break
8:    **else**
9:       Continue
10:   **end if**
11: **end for**
12: **Return** m
13:
14: **Log Recovery:**
15: $g_0 = 2$
16: Given decrypted plaintext $m' = g_0^m \pmod{p}$, $g_0^m \leq p$
17: $m = \log_{g_0}(g_0^m)$
18: **Return** m

4.2.3 Ternary Quantization and Approximate Aggregation

It is unrealistic to encrypt or decrypt all model parameters communicated between the server and clients. If parallel computing is not considered, the computational overhead of model encryption is proportional to the number of model parameters, making it computationally prohibitive for large ML models such as deep neural networks (DNNs). On the other hand, encrypted model parameters occupy much more communication resources due to setting large key length for security consideration. Consequently, ternary gradient quantization and approximate aggregation strategies are used here to deal with these two issues.

Ternary gradient quantization (TernGrad) [57] is originally proposed to reduce the communication costs in distributed deep learning, which is also well suited for encryption in FL. The core idea of TernGrad is to decompose the original full precision gradient tensor into a real value s_t and a ternary precision gradient tensor including only three integers $\{-1, 0, 1\}$, as shown in Eq. (4.9).

$$\tilde{g}_t = s_t \cdot \text{sign}(g_t) \cdot b_t$$
$$s_t = \max(\text{abs}(g_t)),$$

$$(4.9)$$

where g_t is the full precision gradient tensor at the t-th iteration, $sign$ transfers g_t into binary precision with values $\in \{-1, 1\}$, s_t (a scalar) is the maximum absolute element value of g_t and b_t is a binary precision tensor with element values $\in \{0, 1\}$. Consequently, the element-wise product $sign(g_t) \cdot b_t$ is a ternary precision tensor. The binary tensor b_t follows the Bernoulli distribution [58] as described in Eq. (4.10).

$$Pr(b_{tk} = 1|g_t) = |g_{tk}|/s_t$$
$$Pr(b_{tk} = 0|g_t) = 1 - |g_{tk}|/s_t \qquad (4.10)$$

where b_{tk} and g_{tk} is the k-th element of b_t and g_t, respectively. From above Eq. (4.10), it is easy to find that elements in b_t with large absolute values are more likely to be transferred into 1, and vice versa. Furthermore, both g_t and s_t are random variables computed by the input data x_t and the training model parameters θ_t. Thus, Eq. (4.10) can also be re-written into Eq. (4.11).

$$Pr(b_{tk} = 1|x_t, \theta_t) = |g_{tk}|/s_t$$
$$Pr(b_{tk} = 0|x_t, \theta_t) = 1 - |g_{tk}|/s_t \qquad (4.11)$$

And then, the unbiasedness of TernGrad algorithm can be easily proved in Eq. (4.12).

$$\begin{aligned}\mathbb{E}(s_t \cdot sign(g_t) \cdot b_t) &= \mathbb{E}(s_t \cdot sign(g_t) \cdot \mathbb{E}(b_t|x_t)) \\ &= \mathbb{E}(s_t \cdot sign(g_t) \cdot |g_{tk}|/s_t) \qquad (4.12) \\ &= \mathbb{E}(g_t)\end{aligned}$$

In distributed deep learning (data parallelism [59]), the calculated model gradients on each local device should be sent to the parameter server for gradients aggregation and model update. In this case, the TernGrad algorithm can decompose the original model gradients g_t into a positive scalar s_t and a ternary model gradients $sign(g_t) \cdot b_t$. It should be emphasized that the element size of model gradients g_t is always 32 bits (64 bits for double precision), which is 16 times larger than that of ternary gradients $sign(g_t) \cdot b_t$ with only 2 bits.

This phenomenon becomes more obvious in distributed encryption in HFL, since the element size of encrypted model gradients $Enc(g_t)$ is equal to the key length, which is always set to be very large (e.g. 1024 bits) for security considerations. And uploading this kind of gradient ciphertexts $Enc(g_t)$ to the central server consumes a large amount of communication resources. Fortunately, TernGrad can effectively alleviate transferring large gradient ciphertexts by only encrypting the decomposed scalar s_t. Consequently, each client just uploads an encrypted scalar $Enc(s_t)$ and a ternary precision tensor for each layer, producing negligible extra communication costs after encryption. By the way, this approach can also significantly enhance the encryption speed by encrypting only one scalar value s_t for each DNN layer, making HE applicable to federated deep learning.

It is unnecessary to encrypt the ternary gradients containing parts of gradients' sign information. And existing popular attack methods [19, 29, 60–64] performed on FL cannot retrieve the original training data from the ternary gradients. Another advantage of using TernGrad in our proposed federated encryption scheme is that it avoids brute force recovery trying negative values, since all the generated s_t are positive.

Although ternary quantization can effectively reduce both communication costs and local computational overhead on federated encryption, it does no help for decryption. When the server receives the encrypted s_t and their corresponding ternary gradients from connected clients, it needs to multiply these two parts into several encrypted model gradients for the global model aggregation as shown in Eq. (4.13).

$$\text{Enc}(g_t^{\text{global}}) = \sum_i \frac{n_i}{n} (\text{Enc}(s_t^i) * g_{(t,\text{tern})}^i) \tag{4.13}$$

where $g_{(t,\text{tern})}^i$ is the ternary gradients for client i at the t-th communication round, n_i is the data size of client i and n represents the total data size. It is obvious to see that decrypting this aggregated global gradients $\text{Enc}(g_t^{\text{global}})$ is computationally prohibitive by traversing each single ciphertext. Even worse, the global encrypted gradients $\text{Enc}(g_t^{\text{global}})$ requires to be downloaded to at least T (the threshold value of Shamir Secret Sharing) qualified clients for partial federated decryption. And T partial decrypted gradients are still ciphertexts, which should be uploaded to the server for further decryption aggregation. After that, the transformed plaintexts of the global gradients can be finally derived. Therefore, the total number of additional communication costs consumed for distributed decryption is equal to $2T$ times the size of encrypted model gradients. Since the encrypted gradients of DNNs often contain millions of ciphertexts with a large bit length, the communication overhead of distributed decryption is a big burden to FL systems.

In order to simultaneously reduce both computational and communication costs caused by federated decryption, an approximate aggregation strategy is proposed by separately aggregating the encrypted scalar s_t and ternary precision gradients $g_{(t,\text{tern})}$, as described in Eq. (4.14).

$$\text{Enc}(s_t^{\text{global}}) = \prod_i \text{Enc}\left(s_t^i * \frac{n_i}{n}\right) = \text{Enc}\left(\sum_i s_t^i * \frac{n_i}{n}\right)$$
$$g_{(t,\text{tern})}^{\text{global}} = \sum_i g_{(t,\text{tern})}^i \tag{4.14}$$

where $g_{(t,\text{tern})}^{\text{global}}$ is the aggregated global ternary gradients and $\text{Enc}(s_t^{\text{global}})$ is the aggregated encrypted scalar value decomposed by TernGrad. The process of approximate aggregation algorithm is also shown in Fig. 4.1.

After the encrypted scalars $\text{Enc}(s_t^{\text{global}})$ and ternary gradients $g_{(t,\text{tern})}^{\text{global}}$ are generated on the server, only $\text{Enc}(s_t^{\text{global}})$ are required to be downloaded to T or more clients for partial decryption. And then, the partial decrypted ciphertexts pd_t on each client will be returned to the server for decryption aggregation. Finally, the approximated

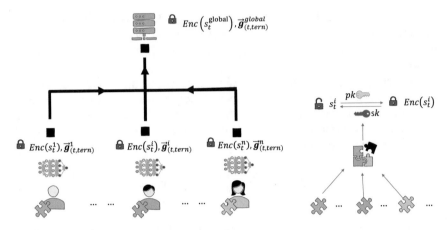

Fig. 4.1 Approximate model aggregation

global gradients is integrated by multiplying the decrypted s_t^{global} and the global ternary gradients $g_{(t,\text{tern})}^{\text{global}}$. Consequently, only the encrypted scalars $\text{Enc}(s_t^{\text{global}})$ (or partially decrypted scalars pd_t) instead of the entire gradients $\text{Enc}(g_t^{\text{global}})$ are needed to be transferred and decrypted, thereby dramatically decreasing the communication costs consumed for distributed decryption and local computational costs for partial decryption.

However, the approximated global gradients $s_t^{\text{global}} * g_{(t,\text{tern})}^{\text{global}}$ is not equal to the actual global gradients g_t^{global}, and that is the reason why this proposed algorithm is called approximate aggregation. The bias $g_t^{\text{global}} - s_t^{\text{global}} * g_{(t,\text{tern})}^{\text{global}}$ becomes zero only if $s_t^1 = s_t^2 = \cdots = s_t^n$. In reality, each client's local scalar satisfies $s_t^1 \approx s_t^2 \approx \cdots \approx s_t^n$ as empirically proved in the experiments, and the bias is negligible in this case.

4.2.4   Overall Framework

We combine the methods introduced in Sects. 4.2.2 and 4.2.3 into a distributed additive ElGamal encryption and quantization for the privacy-preserving federated deep learning framework, DAEQ-FL for short. The detailed steps of DAEQ-FL are described in Algorithm 3.

In the algorithm, FKG is performed at the beginning of each communication round to generate the global key pairs and two verifiable secret sharing methods are also adopted to determine the qualified (QUAL) clients. If the number of QUAL clients are less than T (the threshold value), the disqualified clients need to be removed from the FL system and FKG will be restarted. After FKG, the global model θ_t should be downloaded to each QUAL client i for local training. And then, each

---

Algorithm 3 DAEQ-FL

1: **Server:**
2: Generate and distribute p, q, g, y and global model parameters θ_0
3: **for** each FL round $t = 1, 2, ...$ **do**
4:     Select $n = C \times N$ clients, $C \in (0, 1)$
5:     Select $T > n/2$
6:     Generate pk by FKG among n clients in Algorithm 14
7:     **for** each client $i \in Qual$ in parallel **do**
8:         Download θ_t
9:         Do local **Training**
10:        Upload $c^i_{(t,1)}, c^i_{(t,2)}$ and $\Delta\theta^i_{(t,tern)}$
11:    **end for**
12:    $c_{(t,1)} = \prod_i c^i_{(t,1)}$ (mod p)
13:    $c_{(t,2)} = \prod_i c^i_{(t,2)}$ (mod p)
14:    $\Delta\theta_{(t,tern)} = \sum_i \Delta\theta^i_{(t,tern)}$
15:    Randomly select T $Qual$ clients
16:    **for** each client $j \in T$ in parallel **do**
17:        Download $c_{(t,1)}$ and $c_{(t,2)}$
18:        Do **Partial Decryption**
19:    **end for**
20:    $g_0^{Tm_t} = \prod_{j \in T} pd_j = g_0^{Tm_t} g^{(x - \sum_j \lambda_j x_i)T}$ (mod p)
21:    Recover Tm_t by Algorithm 15
22:    $\theta_{t+1} = \theta_t - \Delta\theta_{(t,tern)} * Tm_t/T$
23: **end for**
24:
25: **Client i:**
26: // **Training:**
27: $\theta^i_t = \theta_t$
28: **for** each iteration from 1 to E **do**
29:    **for** batch $b \in B$ **do**
30:        $\theta^i_t = \theta^i_t - \eta\nabla L_i(\theta^i_t, b)$
31:    **end for**
32: **end for**
33: $\Delta\theta^i_t = \theta^i_t - \theta_t$
34: Quantize $\Delta\theta^i_t$ into s^i_t and $\Delta\theta^i_{(t,tern)}$
35: Encode $m^i_t = round(s^i_t * D_k/D * 2^l)$ (mod q)
36: Encrypt $c^i_{(t,1)} = g^{r_i}$ (mod p), $c^i_{(t,2)} = g_0^{m^i_t} pk^{r_i}$ (mod p)   ▷ r_i is a random number, $r_i \in \mathbb{Z}^*_q$
37: **Return** $c^i_{(t,1)}, c^i_{(t,2)}$ and $\Delta\theta^i_{(t,tern)}$ to server
38: // **Partial Decryption:**
39: $x_i = \sum_j s_{ji} = f_j(i), i, j \in QUAL$
40: Partial decrypt $pd_i = c_{(t,2)}/c^{\lambda_i x_i T}_{(t,1)}$ (mod p)
41: **Return** pd_i

---

updated model gradients $\Delta\theta^i_t$ will be decomposed into a real-valued scalar s^i_t and the corresponding ternary gradient tensor $\Delta\theta^i_{(t,\text{tern})}$. Each scalar s^i_t should be encrypted into two ciphertexts $c^i_{(t,1)}$ and $c^i_{(t,2)}$ by the proposed additive ElGamal before being uploaded to the server together with $\Delta\theta^i_{(t,\text{tern})}$ for approximate aggregation, as shown from line 12 to line 14 in Algorithm 3.

After that, only two aggregated ciphertexts $c_{(t,1)}$ and $c_{(t,2)}$ are required to be downloaded to at least T QUAL clients for partial decryption (from line 38 to line 41 in Algorithm 3). And each client i uploads the partially decrypted ciphertext pd_i to the server for decryption aggregation to get the transformed plaintext $g_0^{Tm_t}$, as shown at line 21 of Algorithm 3. The equation $g_0^{Tm_t} = g_0^{Tm_t} g^{(x-\sum_j \lambda_j x_i)T}$ (mod p) holds only if $x - \sum_j \lambda_j x_i = 0$, which can be easily proven in Eq. (4.15).

$$
\begin{aligned}
\sum_{i \in T} \lambda_i x_i &= \sum_{i \in T} \lambda_i \sum_{j \in \text{QUAL}} f_j(i) \\
&= \sum_{j \in \text{QUAL}} \sum_{i \in T} \lambda_i f_j(i) \\
&= \sum_{j \in \text{QUAL}} z_j = x
\end{aligned}
\tag{4.15}
$$

where $s_{ji} = f_j(i)$ is a share on client j sent to client i as introduced in Eq. (4.1) of Sect. 4.2.2.2, $\lambda_i = \prod_{j \neq i} \frac{j}{j-i}$ is the Lagrange coefficient and $x = \sum_{j \in \text{QUAL}} z_j$ is the global private key.

Finally, the decrypted $g_0^{Tm_t}$ should be converted into the desired plaintext Tm_t/T by brute force or log recovery before multiplying with the aggregated global ternary gradients $\Delta \theta_{(t,\text{tern})}$ to generate the approximated gradients. And the global model for the next communication round can be updated by equation $\theta_{t+1} = \theta_t - \Delta \theta_{(t,\text{tern})} * Tm_t/T$ at line 22 of Algorithm 3.

It is clear to see that both the central server and clients do not have any knowledge of the global private key x, effectively enhancing the security level of the FL system. The proposed federated decryption scheme is robust to client disconnection issues, since only T QUAL clients complete partial decryption operations and upload their pd_i to the server, the ciphertexts can be successfully decrypted. However, the decrypted plaintexts with a large encoding length should be converted into the desired content by brute force, which is very time consuming. And log recovery can only be used in the cases when the condition $g_0^m < G_s(p)$ satisfies, where $G_s(p)$ is the group size of large prime number p. This extremely restricts the encoding bit size of the plaintexts, and may cause unexpected model performance bias during the training. The above mentioned issue will be further discussed in the empirical studies.

4.2.5 Empirical Studies

In this section, we first introduce the datasets and models used in the experiments, together with all settings of the FL and encryption. We also present the communication and computation cost resulting from encryption, as well as the time consumption of the brute force recovery, followed by the analysis of approximate aggregation and a description of results on the model performances.

4.2.5.1   Dataset and Model

Three datasets MNIST, CIFAR10 and Shakespeare are adopted in our experiments. MNIST and CIFAR10 are image datasets used for image classification tasks, and Shakespeare is a text dataset used for next word prediction task.

MNIST is a 28×28 grey scale digit number image dataset containing 60,000 training images and 10,000 testing images with 10 different kinds of label classes (0~9). All the clients' training data are distributed according to their label classes and most clients contain only two kinds of digits for Non-IID data partitions.

CIFAR10 contains 10 different kinds of 50,000 training and 10,000 testing $32 \times 32 \times 3$ images. Similar to MNIST, the whole training data are horizontally sampled and each client owns five different classes of object images.

The Shakespeare dataset is built from the whole work of Williams Shakespeare. It has in total 4,226,073 samples with 1129 role players, which represent users or clients in the FL system. And 90% of the user's data are randomly partitioned as the training data and the rest 10% are testing data. This dataset is naturally Non-IID and unbalanced, with some clients having few lines and others a large number of lines. Furthermore, only 5% of users are randomly selected for training, following the method proposed in [65].

A CNN model is applied to train the MNIST dataset, which contains two 3×3 convolution layers with 32 and 64 filter channels, respectively, followed by a 2×2 max pooling layer and a fully connected layer with 128 neurons. And a ResNet18 is adopted to train on the CIFAR10 dataset, containing four sequentially connected block layers with 64, 128, 256, 512 filter channels, respectively. Besides, the trainable model parameters of batch normalization layers are disabled for simplicity. The numbers of model parameters for CNN and ResNet18 are 1,625,866 and 11,164,362, respectively. Finally the Shakespeare dataset is trained by a stacked LSTM model containing two LSTM layers, each with 256 neurons. And the proposed LSTM model contains 819,920 parameters.

4.2.5.2   Settings of Encryption and Federated Learning

The total number of communication rounds used in FL is set to be 200 for image classifications, and 100 for next word prediction. The number of participants in each communication round is 20 for the MNIST and CIFAR10 dataset, and 10 out of 36 clients for the Shakespeare dataset.

The conventional SGD algorithm is adopted for local client training, and for the CNN models, the number of local epochs is set to 2, the batch size is 50, and the learning rate is 0.1 with a decay rate of 0.995. We do not use any momentum for training the CNNs, while the momentum is set to be 0.5 for the ResNet. For the LSTM, the local epoch is set to 1, the batch size is 10, and the learning rate is 0.5 with a decay rate of 0.995.

The threshold value T is set to be $0.6n$ ($T > n/2$), where n is the number of connected clients in each communication round. The key size and group size of the distributed additive ElGamal encryption are set to 256 and 3072, respectively, to offer a 128-bit security level, and the bit length b for encoding is chosen to be from 2 to 15. In order to reduce the computational overhead of the plaintext recovery, g_0 is set to be 2. And log recovery can only be adopted when the encoded message m satisfies the condition $2^m \leq 3072$ and the corresponding encoding bit length $b \leq 11.6$. Also considering the overflow problem for approximate aggregation, the log recovery is used when the encoding bit length b ranges from 2 to 10, while the brute force recovery method is adopted when the bit length is larger than 10.

4.2.5.3   Encryption Cost and Brute Force Recovery Time

Firstly, the average communication costs caused by distributed encryption between DAEQ-FL and the threshold based Paillier method are presented in Table 4.1. It is obvious to see that our proposed method consumes much less communication resources than the Paillier approach. To be more specifically for LSTM, the communication costs of DAEQ-FL for each round is 0.212MB, which is about 4249 times smaller than that of the Paillier based variants.

Next, the encryption speed of ElGamal and Paillier are listed in Table 4.2. And ElGamal is approximately 17 times and 10 times, respectively, faster than Paillier for encryption and decryption. However, the transformed message in additive ElGamal HE consumes extra recovery time to be converted into the desired plaintext. Thus, in the following, the brute force recovery time with different encoding lengths are plotted in Fig. 4.2 for CNN, ResNet18 and stacked LSTM, respectively.

Considering that our computational resources are limited, the encoding bit length of brute force recovery comparison is set from 11 bits to 15 bits. And it is apparent to find that the computational time becomes larger with the increase of the encoding

Table 4.1 Average communication costs for both encryption and partial decryption with a security level of 128 bits, # of ciphertexts means the number of transmitted ciphertexts for encryption and decryption, respectively

Models	Enc uploads (MB)	Dec downloads (MB)	Dec uploads (MB)	# of Ciphertexts
CNN (DAEQ-FL)	0.3876+0.0059	0.0059	0.0029	16+24
ResNet (DAEQ-FL)	2.6618+0.0161	0.0161	0.0081	44+66
LSTM (DAEQ-FL)	0.1955+0.0066	0.0066	0.0033	18+27
CNN (Paillier)	595.4099	595.4099	595.4099	1625866+3251732
ResNet (Paillier)	4088.5115	4088.5115	4088.5115	11164362+22328724
LSTM (Paillier)	300.2637	300.2637	300.2637	819920+1639840

Table 4.2 Runtime for encryption and decryption of one number using ElGamal and Paillier

Algorithm	Enc time (s)	Dec time (s)
ElGamal	0.0029	0.0015
Paillier	0.0501	0.0141

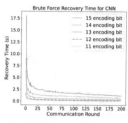 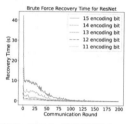 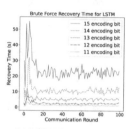

(a) CNN for MNIST dataset   (b) ResNet for CIFAR10 dataset   (c) LSTM for Shakespeare dataset

Fig. 4.2 Brute force recovery time with different encoding bit lengths for different learning models

Table 4.3 Brute force recovery time for 15 encoding bit length

Models	Max (s)	Min (s)	Avg (s)
CNN	18.0467	1.0717	1.9786
ResNet	42.2203	0.0474	2.6491
LSTM	55.2088	12.6166	23.8876

length. Specifically, the CNN and ResNet share similar brute force recovery time, which is very large at the beginning and quickly drops over the communication rounds. This indicates that the gradients of model parameters of SGD decrease as the global model converges. Furthermore, it is surprising to see that the recovery time of the ResNet is almost zero at the end of FL rounds.

Different from the CNN and ResNet, the recovery time of the LSTM does not drop to zero and keeps fluctuating at a relatively high level. There are two possible reasons for this observation. First, the gradients of recurrent connections are accumulated across time sequence, making them much larger than those of the CNN and ResNet. Second, 10 out of 36 clients are randomly selected in the simulations of the LSTM, while all the clients of the CNN and ResNet are participated in federated training. Moreover, the brute force recovery time of 15-bit encoding length for all three models are presented in Table 4.3. And also from Fig. 4.3(a) we can find that the average brute force recovery time accounts for a great proportion of the total elapsed time in each communication round, especially for that of the LSTM.

When the encoding bit length ranges from 2 to 10, the brute force recovery can be replaced by the log recovery to significantly shrink the decryption time. In this

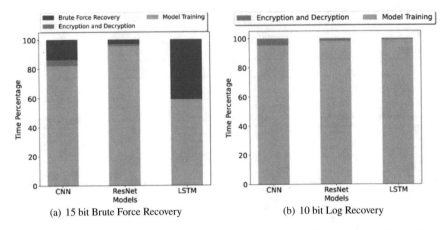

Fig. 4.3 Ratio of consumed time for model training, encryption and decryption, and brute force recovery time

case, as shown in Fig. 4.3(b), the encryption time becomes negligible compared to the total model training time. However, as discussed previously, the log recovery cannot be used when the encoding bit length of plaintext is larger than 11. Moreover, to avoid overflow problems caused by the global model aggregation, the maximum encoding length for adopting the log recovery is set to 10 in the experiments.

4.2.5.4   Analysis of Approximate Model Aggregation

Approximate aggregation analysis is performed by varying the degrees of data skew among connected clients. And three different data partition strategies are used in the experiments. (1) IID, where the training data is randomly allocated to the clients; (2) Non-IID with 5 label classes, where each client only holds five different label classes of the training data; and (3) Non-IID with 2 label classes, in which each client owns two types of the training data. And ResNet is selected for this analysis due to its model complexity with the deepest structure. In addition, for brevity, the maximum absolute elements s_t of the first convolutional layer and the last fully connected layer with a bit encoding length 10 and 15 are presented in Figs. 4.4, 4.5, and 4.6, respectively.

According to aforementioned analysis, the performance bias caused by approximate aggregation can be eliminated only if the condition $s_t^1 = s_t^2 = \cdots = s_t^n$ is satisfied. Overall, it is clear to see that s_t values do not show big differences with each other over the communication rounds. To be more specifically, for IID scenarios, the curves of s_t values are nearly located at a single line in Fig. 4.4, which indicates that s_t values from different clients have almost the same value. In other words, our proposed approximate aggregation technique has similar model performance on IID client data.

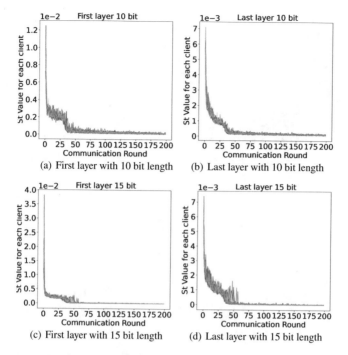

Fig. 4.4 s_t values of all connected clients with IID data

For the Non-IID data with 5 types of the training data, however, the curves show more fluctuations than those of for the IID data. And the observed maximum differences among s_t values are less than 0.003 even for the cases of 10 bit encoding length, which means the approximated bias can also be negligible here. Besides, the results of Non-IID data with only 2 types of local training images are shown in Fig. 4.6. And a clear discrepancy can be found for s_t values of the last fully connected layer and the maximum distance between two s_t values is approximately 0.005. Thus, even for this extreme case, our proposed approximate aggregation algorithm will not bring about large performance biases and still works well in the DEAQ-FL algorithm.

4.2.5.5   Learning Performance

The model performances of DAEQ-FL affected by the TernGrad quantization, approximate aggregation and encoding length are examined and discussed in this section.

The test accuracies of all three cases with and without ElGamal encryption are shown in Fig. 4.7. For non encryption cases, 'Original' represents standard FL, 'Tern-Grad' means only quantization is used and 'TernGrad+Approx' uses both quantization and approximated aggregation techniques. And for encryption cases, brute force

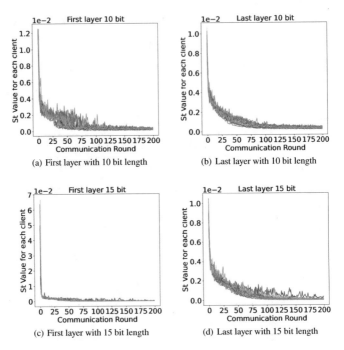

Fig. 4.5 s_t values of all connected clients with Non-IID data where each client contains only 5 different classes of object images

recovery is used for an encoding bit length of 15 and the log recovery is adopted for scenarios with an encoding length of 10.

It is surprising to find that the global model shares almost the same performance among the standard FL and four variants of the proposed DAEQ-FL, which empirically indicates that TernGrad and approximate aggregation have negligible side-effect on the global model performance. Furthermore, two classification tasks of the CNN and ResNet converge at approximately the 25th round, while the LSTM is convergent at about the 50th round and keeps fluctuations afterwards.

The learning performance variations related to the encoding bit length are presented in Fig. 4.8. For two image classification problems, the global model performance starts to degrade when the encoding bit length is less than 9 bits. And the model test accuracy has a dramatic drop when the encoding length is reduced to 7 bits or lower for CNN and to 8 bits or lower for the ResNet. For the next word prediction task, it is surprising to see that reducing the encoding bit length has a minor effect only on the model accuracy of the LSTM until the encoding length is reduced to 2 with 43.15% accuracy. The possible reason is due the large accumulated recurrent gradients.

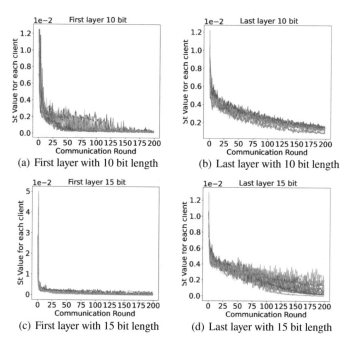

Fig. 4.6 s_t values of all connected clients with Non-IID data where each client contains only 2 different classes of object images

4.3   Secure Vertical Federated Learning

Vertical federated learning (VFL) has completely different training mechanisms and system topology compared to aforementioned horizontal federated learning (HFL). Most VFL systems do not have a central server for model aggregation like HFL systems do, and include a guest client containing all data labels and several host clients without any data labels. A general method for training parametric models in VFL works as in the following. Each host client constructs its local model and uploads model's output other than the model itself to the guest client. And then, the guest client aggregates the received outputs and its own model output to build the total loss function for gradient calculations. Finally, the calculated intermediate gradients will be sent to all connected host clients for local model update. Consequently, model outputs and intermediate gradients are communicated between the guest client and host clients, and the guest client becomes the actual server in VFL. Unlike aforementioned algorithms concentrate on privately training parametric models in HFL, in this section, however, we focus on training non-parametric models in VFL with privacy preservation concerns.

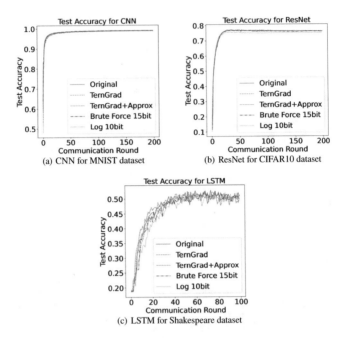

(a) CNN for MNIST dataset

(b) ResNet for CIFAR10 dataset

(c) LSTM for Shakespeare dataset

Fig. 4.7 The test accuracy of the global model for CNN, ResNet and LSTM in five different settings

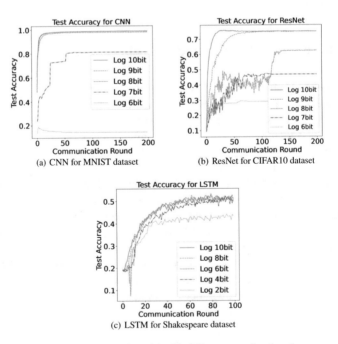

(a) CNN for MNIST dataset

(b) ResNet for CIFAR10 dataset

(c) LSTM for Shakespeare dataset

Fig. 4.8 The test accuracy of the global model with different encoding lengths

4.3.1 *Vertical Federated Learning with XGBoost*

Non-parametric models like gradient boosting decision trees (GBDTs) have been commonly used in FL for vertically partitioned data. XGBoost [66] is a widely used GBDT model in tabular data training because of its better interpretation, easier parameters tuning and faster training process compared with deep neural network models.

Firstly, we will introduce some basic principles of training XGBoost in the centralized learning environment. Consider a training set with m data entries consisting of the feature space $\mathcal{X} = \{x_1, \ldots, x_m\}$ and label space $\mathcal{Y} = \{y_1, \ldots, y_m\}$. Gradients and Hessian values should be calculated to perform node splits and leaf value computations in XGBoost. An example of gradient and Hessian computation for a simple binary classification tree with a sigmoid activation function is shown in Eqs. (4.16) and (4.17), where $\hat{y}_i^{(t-1)}$ denotes prediction of the previous boosting tree (if the current booster tree is denoted as t) for the ith data point. And if the current booster tree is the first one, the prediction $y_i^{(t-1)}$ is always set to be a constant value like 0.5.

$$g_i = \frac{1}{1 + e^{-y_i^{(t-1)}}} - y_i = \hat{y}_i - y_i \tag{4.16}$$

$$h_i = \frac{e^{-y_i^{(t-1)}}}{(1 + e^{-y_i^{(t-1)}})^2} \tag{4.17}$$

If the depth of the current booster tree does not reach the predefined maximum value, the tree node should be split according to all possible impurity scores L_{split} in the predefined buckets shown in Eq. (4.18), where I_L and I_R represent the indices belonging to the left and right branch of the correspond split, respectively, and λ and γ are regularization parameters.

$$L_{split} = \frac{1}{2}\left[\frac{(\sum_{i \in I_L} g_i)^2}{\sum_{i \in I_L} h_i + \lambda} + \frac{(\sum_{i \in I_R} g_i)^2}{\sum_{i \in I_R} h_i + \lambda} - \frac{(\sum_{i \in I} g_i)^2}{\sum_{i \in I} h_i + \lambda}\right] - \gamma \tag{4.18}$$

On the other hand, if the stop criterion is satisfied, the current tree node will be regarded as leaf whose weight value is computed based on Eq. (4.19).

$$w = -\frac{\sum_{i \in I} g_i}{\sum_{i \in I} h_i + \lambda}. \tag{4.19}$$

Label prediction $\hat{y}_i^t$ for each data entry i should be updated before constructing the current decision tree $\mathcal{T}$, and the update rule is presented in Eq. (4.20).

$$\hat{y}_i^t = \hat{y}_i^0 + \sum_{t=1}^{\mathcal{T}-1} \eta w_i^t \tag{4.20}$$

where $\hat{y}_i^0$ is the original label prediction value and is always set to be 0 for classification tasks and 0.5 for regression tasks, t is the tree index, w_i^t is the corresponding leaf output of tree t for data i and η is the learning rate. And it is easy to find that the leaf output w_i^t of each decision tree t for data i is sequentially accumulated for label prediction update. This is the reason why XGBoost is a boosting rather than a bagging ensemble machine learning algorithm.

Similarly, after all the T boosting trees are successfully constructed, the model prediction for each data sample i can be computed according to the following Eq. (4.21), where σ is the activation function that can be removed in the regression problems.

$$\hat{y}_i = \sigma\left(\hat{y}_i^0 + \sum_{t=1}^{T} \eta w_i^t\right) \tag{4.21}$$

Different from aforementioned centralized learning, much more effort should be made on privacy preservation in constructing XGBoost in VFL. Given that X is the feature space, $\mathcal{Y}$ is the label space and $\mathcal{I}$ is the set of data IDs, the standard VFL can be defined as:

Definition Vertical federated learning: a training set with m data points distributed across n clients, each client c has a subset of data features $X^c = \{X_1^c, \ldots, X_m^c\}$ and sample ids $\mathcal{I}^c = \{I_1, \ldots, I_m\}$ where $c \in \{1, \ldots, n\}$. Only one client $l \in \{1, \ldots, n\}$ contains all the labels $\mathcal{Y} = \{y_1, \ldots, y_m\}$. For any two different clients c, c', they satisfy:

$$X^c \neq X^{c'}, \mathcal{Y}^c = \mathcal{Y}^{c'}, \mathcal{I}^c = \mathcal{I}^{c'}, c \neq c' \tag{4.22}$$

It indicates that client c shares the same data sample ids $\mathcal{I}^c$ with any other client c' but has different data features X^c. It should be noticed that $\mathcal{Y}^c$ is an imaginary data label and is only distributed on the guest client. There are two types of clients in the original VFL: one is the *guest* client or *active* party, the other is the *host* client or *passive* party.

Definition Guest Client: it holds both the features (attributes) X^c and all labels $\mathcal{Y}$. In general, a VFL system contains only one guest client. □

Definition Host Client: it only holds data features X without any data labels. A VFL system always contain multiple host clients □

According to these two definitions, gradient and Hessian values (Eqs. (4.16) and (4.17)) can only be computed on the guest client, since only guest client contains data labels. However, impurity score and leaf weight calculations are collaboratively performed between the guest client and host clients. Therefore, the potential privacy leakage may come from two sources during training period: one is the gradients containing label information, and the other is the split information containing feature information.

The first attempt of constructing XGBoost in VFL was reported in [67], which is called SecureBoost. As shown in Fig. 4.9, only intermediate computations are

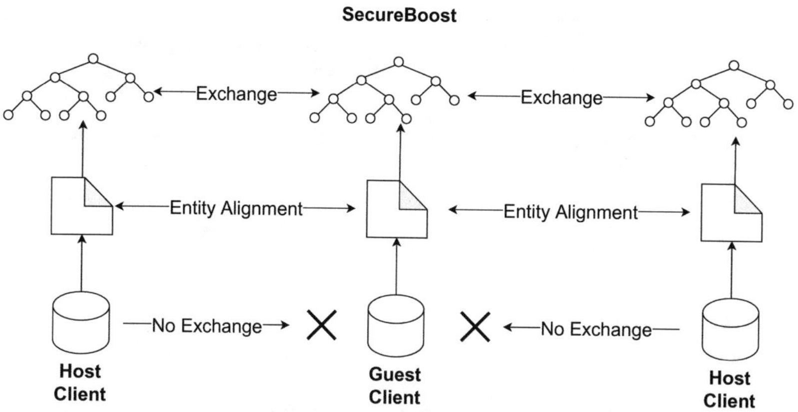

SecureBoost

Fig. 4.9 The secureboost system

allowed to be exchanged during training and local data of each client cannot revealed to any others. In addition, private entity alignment is adopted to find intersected data samples among all connected clients.

In order to privately construct XGBoost in VFL, SecureBoost works in the following way. The guest client calculates the gradients and Hessians of the current boosting tree for all data points and encrypts them into ciphertexts with Paillier HE before sending to other host clients. It is necessary to prevent the gradients from being revealed, since the computed gradients contain label information as mentioned in Eq. (4.16). And then, each host client traverses all possible splits based on the predefined feature thresholds and calculates their corresponding encrypted gradient sums $\text{Enc}(G_L) = \sum_{i \in I_L} \text{Enc}(g_i)$ and Hessian sums $\text{Enc}(H_L) = \sum_{i \in I_L} \text{Enc}(h_i)$ of the left branch. After that, these $\text{Enc}(G_L)$ and $\text{Enc}(H_L)$ together with the record IDs (any integer numbers representing IDs of all possible splits defined by local clients) will be sent back to the guest client for decryption. Finally, the impurity scores L_{split} and leaf weights w for all the splits can be easily calculated with decrypted G_L and H_L on the guest client.

The split with the maximum L_{split} will be regarded as the best split for the current tree node wherein the decision rule is marked with a client ID and a record ID as shown in Fig. 4.10. It should be emphasized that the constructed local lookup table cannot be exchanged, since both data features and split values contain private data information. In addition, all the leaf weights are calculated and stored on the guest client, preventing data labels being deduced from leaf values.

Consequently, during the training period in SecureBoost, each host client is not able to deduce both gradients and Hessians from the received ciphertexts, and the guest client only receive user-defined record IDs without any information of data features and the corresponding split values. Thus, the local private data are effectively protected. Furthermore, for federated inference on the constructed boosting trees, all the clients will collaborate to make the predictions. Any test data should be sent to

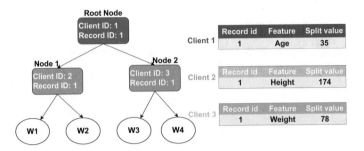

Fig. 4.10 Local lookup table

the correct client to find the corresponding data feature and split value in the local lookup table for decision making according to the client and record IDs within each tree node. And final data predictions in Eq. (4.21) are all computed on the guest client to prevent label leakage.

However, empirical results of SecureBoost indicate that encryption computations occupy a large amount of computational resources. In order to alleviate this issue, Tian et al. propose a more efficient FederBoost framework [68] without encryption operations. Unlike the guest client in SecureBoost directly sending encrypted gradients and Hessians to host clients for encrypted summations, FederBoost enables each host client to sort its own data features into buckets and send data IDs to the guest client. And then, the guest client can calculate the maximum impurity score L_{split} according to data IDs in each bucket. The rest of training and federated inference parts in FederBoost are similar to those in SecureBoost. It is clear to see that only data IDs in each bucket are sent from the host clients to the guest client, while the gradients, Hessians, data labels and label predictions are all stored on the guest client without being shared to any other parties. In addition, differential privacy [32] is also applied to further prevent potential information leakage from orders of data IDs. Therefore, the privacy of the training data can also be guaranteed in FederBoost even without any encryption.

Both SecureBoost and FederBoost are built under the assumption that all the data labels are distributed on one guest client, which is unrealistic in many real-world applications. For instance, in real-life medical systems, different hospitals in the same region may have the same group of patients but provide different modalities of disease tests. And it is very likely that each hospital just owns parts of patients–diagnosis results.

Consequently, in this work, we propose a privacy preserving vertical federated XGBoost system named PIVODL with training labels distributed across multiple clients. The definition of VFL over distributed labels (VFL-DL) is shown below:

Definition VFL-DL Given a training set with m data points, each participating client c consists of data features $\mathcal{X}^c = \{X_1^c, \ldots, X_m^c \mid c \in \{1, \ldots, n\}\}$ and parts of labels $\mathcal{Y}^c = \{y_i^c \mid i \in \{1, \ldots, m\}, c \in \{1, \ldots, n\}\}$. For any two clients $c, c' \in \{1, \ldots, n\}$:

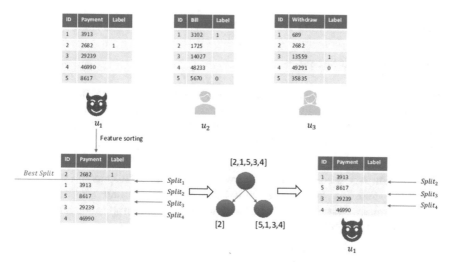

Fig. 4.11 A simple example of split leakage

$$X^c \neq X^{c'}, y^c \neq y^{c'}, I^c = I^{c'}, \forall c \neq c' \tag{4.23}$$

It should be emphasized that in VFL-DL, host clients do not exist, since all participants have parts of data labels. This makes privacy protection much harder than the original VFL in that each client becomes a 'guest' client for both gradient and impurity calculations, causing potential node split and prediction update leakage during the period of tree constructions.

Split leakage comes from two sources. One is the summation differences between all possible splits of the same tree node, and the other is the summation differences between the parent and child nodes. A simple example is shown in Fig. 4.11, where all the splits from $Split_1$ to $Split_4$ are calculated. For example, on client u_1 with the help of u_2 and u_3 for node split. If u_1 is adversarial, the label information on u_2 and u_3 can be easily deduced by *differential attack*, even though the gradients and Hessians are encrypted. For instance, client u_1 requires gradient sums $G_{1,split_2} = g_2 + g_1$ and $G_{1,split_3} = g_2 + g_1 + g_5$ to calculate L_{split_2} and L_{split_3}, respectively. And g_5 from u_2 can be directly computed by $G_{1,split_3} - G_{1,split_2}$ no matter whether g_5 is encrypted or not.

Prediction update leakage is caused by updating label predictions once a new boosting tree is constructed. As introduced previously, XGBoost is an ensemble boosting ML algorithm that requires to sequentially build multiple decision trees with prediction updates for the learning task. And label prediction $\hat{y}_i^t$ of each data point is upgraded by the corresponding leaf weight (Eq. (4.20)) performed on the guest client of the original VFL system. However, in VFL-DL, all the participating clients hold parts of data labels and it is likely that some clients receive some of leaf values of other clients for prediction updates. Consequently, the actual data labels may be guessed or inferred from these received leaf weights.

4.3.2   Secure Node Split and Construction

4.3.2.1   Secure Node Split

To solve the node split leakage problem, secure node split protocol is proposed by separately performing feature splits and summation aggregations on two different types of clients named *source client* and *split client*, respectively. In the following, we define the roles of both source and split clients.

Definition Source client is the split owner that contains all data IDs of the local features. However, it does not have any information of the corresponding summation values $G_{j,v}$ and $H_{j,v}$, where j is the index of data features and v is the index of threshold values. Therefore, it is not able to calculate the impurity scores in Eq. (4.18) and leaf weights in Eq. (4.19).

Definition Split client, on the other hand, is used to compute both the impurity scores in Eq. (4.18) and leaf weights in Eq. (4.19), however, it has no idea of their corresponding data IDs on other clients.

By utilizing both source and split clients during node split, the data labels can be effectively protected without being leaked. The secure node split protocol can be divided into the following steps:

1. Before federated training starts, each connected participating client locally generates its key pairs of Paillier. The generated public key would be shared to other parties while the private key is secretly stored on its own local device. In addition, each client exchanges data IDs of the training samples containing labels I^c, where c is the index of client ID.
2. For root node split, the total gradient sum G and Hessian sum H of all the training samples should be calculated at first without privacy leakage. And these two sums are privately computed on two random clients: one is *encryption client* for decryption operations and the other is *aggregation client* for sum aggregation. Each client c computes it own gradient sum $G^c = \sum_i g_i^c$ and Hessian sum $H^c = \sum_i h_i^c$ and encrypts them by the public key of the encryption client. And then the encrypted two sums $\langle G^c \rangle$ and $\langle H^c \rangle$ of each client c will be sent to the aggregation client to compute the total summation $\langle G \rangle = \prod_c \langle G^c \rangle = \langle \sum_c G^c \rangle$ and $\langle H \rangle = \prod_c \langle H^c \rangle = \langle \sum_c H^c \rangle$, respectively. After that, the aggregation client sends $\langle G \rangle$ and $\langle H \rangle$ to the encryption client for decryption: $G = Dec \langle G \rangle$, $H = Dec \langle H \rangle$.
3. After both G and H of the current tree node are broadcast, each client is performed as the source client to compute gradient sum $G_{j,v}^c = \sum_{i \in \{i | x_{i,j} < v\}} g_i^c$ and Hessian sum $H_{j,v}^c = \sum_{i \in \{i | x_{i,j} < v\}} h_i^c$ of the left branch for all possible splits according to the available local labels $y_{i,j}^c$. In general, $G_{j,v}^c \neq G_{j,v}$ and $H_{j,v}^c \neq H_{j,v}$ unless client c contains all the corresponding labels of the left branch for the split with feature j and threshold v. Therefore, each client c should randomly select any one client as the split client and sends the intersected missing data IDs $I_{j,v}^{c'} = I_{j,v}^c \bigcap I^{c'}$, $c' \in [1, n]$, $c' \neq c$ together with its split client ID and temporary record number $R_{j,v}$ (user defined split ID) to other connected clients.

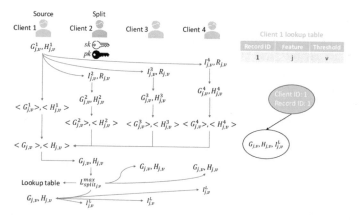

Fig. 4.12 Secure node split for the left branch, where we assume the feature j and threshold v will give the best split with the largest impurity score L_{split}

4. After receiving intersected data IDs and its split client ID from source clients, each client c calculates $G_{j,v}^c = \sum_{i \in \{i | i \in I_{j,v}^c\}} g_i^c$ and $H_{j,v}^c = \sum_{i \in \{i | i \in I_{j,v}^c\}} h_i^c$, and uses the public key to encrypt the sum values according to the received split client ID. And then, two encrypted sums $\langle G_{j,v}^c \rangle$ and $\langle H_{j,v}^c \rangle$ will be sent back to their source client to compute the total sums $\langle G_{j,v} \rangle = \prod_c \langle G_{j,v}^c \rangle = \langle \sum_c G_{j,v}^c \rangle$ and $\langle H_{j,v} \rangle = \prod_c \langle H_{j,v}^c \rangle = \langle \sum_c H_{j,v}^c \rangle$ of the split.

5. Each source client sends $\langle G_{j,v} \rangle$ and $\langle H_{j,v} \rangle$ to its corresponding split client c' for decryption and the split clients can compute the impurity scores by $L_{split_{j,v}} = \frac{1}{2} \left[\frac{G_{j,v}^2}{H_{j,v}+\lambda} + \frac{(G-G_{j,v})^2}{H-H_{j,v}+\lambda} - \frac{G^2}{H+\lambda} \right] - \gamma$. All the selected split clients get the largest local impurity score at first and then broadcast these values for further comparison to achieve the largest global impurity score $L_{split_{j,v}}^{max}$. The split client that owns $L_{split_{j,v}}^{max}$ would send the temporary record number $R_{j,v}$ back to the source client. The source can get the best split feature j and threshold v according to received $R_{j,v}$ and privately build a lookup table to record j and v with a unique *record ID*. And then the tree node can be constructed by the source client ID and the record ID as shown in Fig. 4.12.

6. If the current tree node is not the leaf node, either the left or right child node will continue splitting. If the left node becomes the current node for splitting, the split client sets $G = G_{j,v}$, $H = H_{j,v}$, and sends them to all other clients. Otherwise, the split client sets $G = G - G_{j,v}$, $H = H - H_{j,v}$, and sends them to all other clients. In addition, the source client sends the left split node data IDs $I_{j,v}^L$ or the right split node data IDs $I_{j,v}^R$ to other clients. Then, each client computes the current information gain $\frac{1}{2} \frac{G^2}{H+\lambda}$. Go back to step (3) and repeat the process.

It is clear to see that the proposed secure node split protocol can effectively prevent the split leakage caused by differential attack. The source client c knows the data IDs of two branches and the corresponding split feature j and threshold value v, but has no idea about its computed gradient sum and Hessian sum. On the other hand, the split client can get two computed sums, but has no knowledge about the split information. Note that, to reduce the potential privacy leakage, each split client computes and achieves the largest local impurity score L_{split} for broadcasting. The maximum impurity score $L_{split_{j,v}}^{max}$ can be derived from broadcast shares and its corresponding sums $G_{j,v}$ and $H_{j,v}$ would be regarded as G and H of the next tree node. Furthermore, the split client c' also require to send the temporary record number $R_{j,v}^{c,best}$ of the best split to the source client c. After that, the source client can share the corresponding $I_{j,v}$ to all participants for continuing split. Consequently, it is possible that an adversarial client c^{adv} can infer the gradients from other clients between the received $G_{j,v}$ and locally computed $G_{j,v}^{c^{adv}}$.

In order to alleviate this privacy leakage between the parent and child nodes, we adopt a simple data instance threshold strategy. For example, if the number of received intersected data IDs is less than the pre-defined instance threshold (e.g. 10), this client will reject to compute the local sum $G_{j,v}^c$ and this split will be automatically regarded as invalid. It should be mentioned that continue splitting is robust to differential attacks, since only one $G_{j,v}$ and $H_{j,v}$ are broadcast to each client. The overall secure node split protocol is also shown in Algorithm 4.

4.3.2.2   Private Node Construction

As mentioned before, the split client c' containing the largest impurity score will send the corresponding temporary record number $R_{j,v}^{c,best}$ to the source client c. And then, the source client c should add the feature j and threshold v with a unique record ID into the local lookup table, as shown in Fig. 4.12. After that, client c broadcasts the unique record ID to any other client that can annotate the split of current tree node with client c's ID and the record ID.

Meanwhile, the source client c can check and tell the split client c' whether the split child nodes are a leaf or not based on the condition shown in line 5 and line 15 of Algorithm 5, respectively. If the child node is not a leaf, the source client c broadcasts the split data IDs and the split client c' broadcasts the corresponding sum of gradients G and Hessian H for continuing split. If the child node is determined to be a leaf node, the split client c' computes the local leaf weight and stores it in a *received* lookup table with a source client c ID and a user defined record ID. Both the lookup table and received lookup table are stored locally without being shared with other clients for privacy concern.

Note that both $G_{j,v}$ and $H_{j,v}$ in the algorithm are the gradient and Hessian sums of the left split branch, and those of the right split branch can be easily calculated by $G - G_{j,v}$ and $H - H_{j,v}$.

The reason for constructing a locally stored lookup table is to prevent split data feature j and threshold v of the source client from being revealed, since the decision rules within each tree node of the boosting trees are just the client ID and record ID. And the shortcoming of this strategy is that all the related clients should collaborate to compute the new instance predictions. As a simple example shown in Fig. 4.13, if

Algorithm 4 Secure node split protocol

1: **Input:** I, data IDs of current tree node
2: **Input:** G, gradient sum of current tree node
3: **Input:** H, Hessian sum of current tree node
4: **Input:** T, instance threshold of current tree node
5:
6: *Split client sets $C' = \emptyset$*
7: **for** each *source client* $c = 1, 2, ...n$ **do**
8:     Randomly select a *split client* c', $c' \in [1, n], c' \neq c$
9:     $L_{split}^{c'} \leftarrow 0, C' \leftarrow C' \cup c'$
10: **end for**
11:
12: **for** each *source client* $c = 1, 2, ...n$ **do**
13:     $R_{j,v}^c \leftarrow 0$
14:     **for** each feature $j = 1, 2, ...d^c$ **do**
15:         **for** each threshold $v = 1, 2, ...b_j$ **do**
16:             $pk \leftarrow pk$ of client c'
17:             $G_{j,v}^c \leftarrow \sum_{i \in \{i | x_{i,j} < v\}} g_i^c, \langle G_{j,v}^c \rangle \leftarrow Enc_{pk}(G_{j,v}^c)$
18:             $H_{j,v}^c \leftarrow \sum_{i \in \{i | x_{i,j} < v\}} h_i^c, \langle H_{j,v}^c \rangle \leftarrow Enc_{pk}(H_{j,v}^c)$
19:             $R_{j,v}^c \leftarrow R_{j,v}^c + 1$
20:             **for** each *client* $c'' = 1, 2, ...n, c'' \neq c$ **do**
21:                 $I_{j,v}^{c''} \leftarrow I_{j,v}^c \bigcap I^{c''}$
22:                 Send $I_{j,v}^{c''}, R_{j,v}^c$ to client c'':
23:                 $pk \leftarrow pk$ of client c'
24:                 **if** $|I_{j,v}^{c''}| \geq T$ **then**
25:                     $G_{j,v}^{c''} \leftarrow \sum_{i \in I_{j,v}^{c''}} g_i^{c''}, \langle G_{j,v}^{c''} \rangle \leftarrow Enc_{pk}(G_{j,v}^{c''})$
26:                     $H_{j,v}^{c''} \leftarrow \sum_{i \in I_{j,v}^{c''}} h_i^{c''}, \langle H_{j,v}^{c''} \rangle \leftarrow Enc_{pk}(H_{j,v}^{c''})$
27:                     Return $\langle G_{j,v}^{c''} \rangle$ and $\langle H_{j,v}^{c''} \rangle$ to *source client* c
28:                 **end if**
29:             **end for**
30:             **if** *Source client* c receives $n - 1$ $\langle G_{j,v}^{c''} \rangle$ and $\langle H_{j,v}^{c''} \rangle$ **then**
31:                 $\langle G_{j,v} \rangle = \prod_c \langle G_{j,v}^c \rangle = \langle \sum_c G_{j,v}^c \rangle$
32:                 $\langle H_{j,v} \rangle = \prod_c \langle H_{j,v}^c \rangle = \langle \sum_c H_{j,v}^c \rangle$
33:                 Send $\langle G_{j,v} \rangle$ and $\langle H_{j,v} \rangle$ to *split client* c':
34:                 $G_{j,v} \leftarrow Dec\langle G_{j,v} \rangle$
35:                 $H_{j,v} \leftarrow Dec\langle H_{j,v} \rangle$
36:                 $L_{split_{j,v}} = \frac{1}{2} [\frac{G_{j,v}^2}{H_{j,v}+\lambda} + \frac{(G-G_{j,v})^2}{H-H_{j,v}+\lambda} - \frac{G^2}{H+\lambda}] - \gamma$
37:                 **if** $L_{split_{j,v}} > L_{split}^{c'}$ **then**
38:                     $L_{split}^{c'} \leftarrow L_{split_{j,v}}, R_{j,v}^{c,best} \leftarrow R_{j,v}^c$
39:                 **end if**
40:             **end if**
41:         **end for**
42:     **end for**
43: **end for**
44:
45: **Output:** the *split client* c' with the largest impurity score $L_{split}^{c'}$

Algorithm 5 Construction of private tree nodes, where c' is the split client with the largest impurity score, c is the source client of c', $R^{c,best}_{j,v}$ is the temporary record number of the best split on c, and T_{sample} is the minimum data samples for each decision node

---

1: c' sends $R^{c,best}_{j,v}$ and split branch to c
2: c adds feature j and threshold v with a unique record ID into the local lookup table
3: c broadcasts the record ID and each client can annotate the split of current tree node with client
      c ID and record ID
4: **if** Split branch is left **then**
5:    **if** Reach the maximum depth or $|I_{j,v}| < T_{sample}$ **then**
6:       c tells c' left split node is the leaf node and stops splitting
7:       $w^L_{j,v} = -\frac{G_{j,v}}{H_{j,v}+\lambda}$
8:       c' records left leaf value $w^L_{j,v}$ into *received* lookup table with record ID and client c ID
9:    **else**
10:       Continue splitting the left child node:
11:       c broadcasts $I_{j,v}$ to all other clients
12:       c' broadcasts $G_{j,v}$ and $H_{j,v}$ to all other clients
13:    **end if**
14: **else if** Split branch is right **then**
15:    **if** Reach the maximum depth or $|I - I_{j,v}| < T_{sample}$ **then**
16:       c tells c' right split node is the leaf node and stops splitting
17:       $w^R_{j,v} = -\frac{G-G_{j,v}}{H-H_{j,v}+\lambda}$
18:       c' records right leaf value $w^R_{j,v}$ into *received* lookup table with a record ID and a client c
      ID
19:    **else**
20:       Continue splitting the right child node:
21:       c broadcasts $I - I_{j,v}$ to all other clients
22:       c' broadcasts $G - G_{j,v}$ and $H - H_{j,v}$ to all other clients
23:    **end if**
24: **end if**

---

a new data sample with age 30 and weight 150 comes for prediction, it will be sent to client 1 for judgement according to the client ID of the root node. The client 1 seeks the split feature and its corresponding threshold in the local lookup table based on the record ID, and determines this instance to the left branch ($30 < 35$). After that, this data is sent to client 2 to get its label prediction 0.41 ($150 > 120$, go to right leaf). Moreover, the secure aggregation scheme proposed in [69] is adopted to prevent possible label inference during predictions for ensemble trees.

4.3.3   Partial Differential Privacy

Since XGBoost is an ensemble tree learning algorithm, the label predictions of each data point should be upgraded before constructing a new boosting tree. Therefore, each split client in VFL-DL requires to send the computed leaf values w to other clients for prediction updates. And there is no doubt that as the training proceeds,

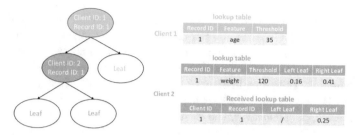

Fig. 4.13 Lookup table on the source client and received lookup table on the split client

the accumulated label prediction $\widehat{y}_i$ will be closer to the true data label y_i. Recall that the source client contains all the split data IDs of local features. Thus, once it receives the leaf weights from the split client, the true labels of the intersected data samples may have the risk of being deduced.

Consequently, a partial differential privacy (DP) mechanism is proposed here to prevent this prediction update leakage. DP [32, 70] is one of data masking techniques originally proposed to protect the privacy of public dataset. The core idea behind DP is that the effect of an arbitrary single variation in the database is small enough to be inferred by the query result. The definition of DP is shown below:

Definition (ϵ-*differential privacy* [70]). Given a real positive number ϵ and a randomized algorithm $\mathcal{A}: \mathcal{D}^n \rightarrow \mathcal{Y}$. Algorithm $\mathcal{A}$ provides ϵ—differential privacy, if for any two datasets D, $D' \in \mathcal{D}^n$ differs on only one entity, and all $\mathcal{S} \subseteq \mathcal{Y}$ satisfy:

$$Pr[\mathcal{A}(D) \in \mathcal{S}] \leq exp(\epsilon) \cdot Pr[\mathcal{A}(D') \in \mathcal{S}] \tag{4.24}$$

The Laplace mechanism is often applied to ϵ-DP by adding Laplacian noise to a deterministic function of a database $\mathcal{D}$. Furthermore, the definition of a weaker form of ϵ, δ-DP called approximated DP [71] is also given as follows:

Definition ((ϵ, δ)-*differential privacy*). Given two real positive numbers (ϵ, δ) and a randomized algorithm $\mathcal{A}: \mathcal{D}^n \rightarrow \mathcal{Y}$. An algorithm $\mathcal{A}$ provides (ϵ, δ)-differential privacy if it satisfies:

$$Pr[\mathcal{A}(D) \in \mathcal{S}] \leq exp(\epsilon) \cdot Pr[\mathcal{A}(D') \in \mathcal{S}] + \delta \tag{4.25}$$

where δ represents the probability of potentially leaking large privacy. And it is easy to find that if $\delta = 0$, (ϵ, δ)-DP is exactly the same as the classical ϵ-DP.

For (ϵ, δ)-DP, the Gaussian mechanism [72–74] instead of the Laplace mechanism is adopted by adding Gaussian noise, $\mathcal{N} \sim N(0, \Delta^2 \sigma^2)$ to the output of the algorithm, where Δ is l_2-norm sensitivity of D and $\sigma \geq \sqrt{2 \ln(1.25/\delta)}$. And this noise can achieve ($O(q\epsilon), q\delta$)-DP, where q is the sampling rate per lot. To make the noise small while satisfying DP, we follow the definition in [72] and set $\sigma \geq c \frac{q\sqrt{T log(1/\delta)}}{\epsilon}$, c is a constant, and T refers to the number of steps, to achieve ($O(q\epsilon\sqrt{T}), \delta$)-DP.

Similarly, our proposed partial DP adopts the Gaussian mechanism by perturbing the leaf value w with Gaussian noise N before sending it to the source client. However, the range of leaf value w is unpredictable since the numbers of summed g and h are not deterministic, and the sensitivity to noise cannot be directly derived from the DP definition. Therefore, in order to solve this issue while simultaneously satisfying (ϵ, δ)-DP requirement, the clipping technique is used here to let the sensitivity equals to the clipping boundary. The sensitivity Δ_w is calculated in Eq. (4.26), where C refers to the predefined clip value and $\{G, H\}, \{G', H'\}$ are two sets, differing in one pair of g and h.

$$
\begin{aligned}
\Delta_w &= \max_{\{G,H\},\{G',H'\}} \|w^{\{G,H\}} - w^{\{G',H'\}}\| \\
&= 2C, \ \forall \|w^{\{G,H\}}\|, \|w^{\{G',H'\}}\| \le C
\end{aligned}
\tag{4.26}
$$

By *partial* DP, we mean that DP is applied on w sent to the source client only, since any other client just knows its intersected local data IDs. The advantage of using this strategy is that the correct distributed label predictions can be achieved to the greatest extent, while preventing the source client from guessing the true labels. The details of the partial DP mechanism is presented in Algorithm 6.

Algorithm 6 Partial differential privacy. c' is the split client and c is the corresponding source client

---
1: **Input:** $w_{j,v}$, computed leaf weight on split client c'
2:
3: **for** each $c'' = 1, 2, ...n, c'' \ne c'$ **do**
4:   **if** $c'' = c$ **then**
5:     $\hat{w_{j,v}} \leftarrow w_{j,v}/max(1, \frac{\|w_{j,v}\|}{C})$
6:     $\hat{w_{j,v}} \leftarrow \hat{w_{j,v}} + N \sim N(0, 4C^2\sigma^2)$
7:     Send perturbed $\hat{w_{j,v}}$ to c
8:   **else**
9:     Send $w_{j,v}$ to c''
10:   **end if**
11: **end for**

4.3.4 Security Analysis

Our proposed PIVODL framework avoids revealing both data features and labels on each participating client. In this section, we will make a detailed security analysis of the leakage sources and protection strategies.

During the training process, parts of g and h from other clients are required to be sent to the source client to find the maximum impurity score L_{split} for node split. The true data labels of other clients can be easily inferred from the gradients g.

Theorem 4.1 *Given a gradient g, the true label of corresponding data point can be inferred.*

Proof According to Eq. (4.16), adversaries are able to deduce the true label of the ith data sample by $y_i = \hat{y}_i - g_i$ if g_i is known.   ☐

Therefore, it is necessary to adopt the encryption to encrypt intermediate $\sum g$ and $\sum h$ for each bucket before sending them to the source client for summation. However, it is possible that the specific gradient information g_i can still be deduced by a differential attack, if the clients have data the ID information of its adjacent bucket splits.

Theorem 4.2 *A potential leakage risk still exists in which the gradient g_i of the ith data sample can be deduced by a differential attack, even if the intermediate $\sum g$ and $\sum h$ are encrypted.*

Proof Assume L_n and $L_{n'}$ are two adjacent data sample sets for two possible splits that only differs in one data sample i (or a few data samples). Although the client only knows the aggregated sum $\sum_{i \in L_n} g_i$ and $\sum_{i \in L_{n'}} g_i$, g_i can be easily derived with a differential attack $g_i = \sum_{i \in L_n} g_i - \sum_{i \in L_{n'}} g_i$.   ☐

Consequently, the source client and the split client are required to perform the bucket split and sum aggregation separately. In this way, the source client only knows the data IDs of the local feature split but has no idea about their gradient and Hessian sum. By contrast, the split client holds summation values but does not know their data IDs.

Before building the next boosting tree, the predictions of all data samples need to be updated based on their leaf weights. It is not surprising that the accumulated leaf weights tend to approach the actual data labels as the training proceeds, thus, attackers can easily guess the data labels with a high confidence level through the received leaf weights.

Theorem 4.3 *Given multiple boosting trees' leaf values, it is likely to deduce the true labels of the corresponding data points.*

Proof A label prediction $\hat{y}_i$ can be calculated with equation $\hat{y}_i = \sigma(\hat{y}_i^0 + \eta \cdot W_i^1 + \cdots + \eta \cdot W_i^t)$, where σ is the activation function and can be removed for the regression problems, η refers to the learning rate and W_i^t is the leaf weight of tth decision tree for data point i. Therefore, once the source client knows parts of or even all leaf values W, the true labels have a high risk of being inferred.   ☐

In order to address this issue, the split client applies the above-mentioned partial DP mechanism before sending the leaf values to the source client.

4.3.5   *Empirical Studies*

In this section, we introduce the experimental settings at first. And then the experimental results will be given, followed by a discussion of the learning performance. Furthermore, the training time of our privacy-preservation method will be evaluated. Finally, the communication cost and label prediction inference of PIVODL will be described.

4.3.5.1   Experimental Settings

Three public datasets are used in our empirical studies, where the first two are for classification tasks and the third is for regression tasks.

Credit card [75]: It is a credit scoring dataset that aims to predict if a person will make payment on time. It contains a total of 30,000 data samples with 23 attributes.

Bank marketing [76]: The data is related to direct marketing campaigns of a Portuguese banking institution. The prediction goal is to evaluate whether the client will subscribe a term deposit. It consists of 45211 instances and 17 attributes.

Appliances e nergy prediction [77]: It is a regression dataset of energy consumption in a low energy building, which has 19735 data instances and 29 attributes.

All these datasets are partitioned into training and test data, each of which occupies 80% and 20% of the entire data samples, respectively. Other experimental settings like the number of participating clients, the number of the maximum depth of the boosting tree are presented in Table 4.4. Furthermore, all the experiments are repeated for five times.

Table 4.4 Experimental settings of PIVODL

Hyperparameters	Range	Default
Number of clients	[2, 4, 6, 8, 10]	4
Maximum depth	[2, 3, 4, 5, 6]	3
Number of trees	[2, 3, 4, 5, 6]	5
Learning rate	/	0.3
Regularization	/	1
Number of buckets	/	32
Encrypt key size	/	512
ϵ	[2, 4, 6, 8, 10]	8
Sensitivity clip	/	2
Sample threshold	/	10

4.3.5.2   Sensitivity Analysis

Here, we empirically analyze the change of the learning performances as the number of participating clients, the maximum depth of the tree structure, the number of trees, and ϵ in the partial DP. The results in terms of the mean, the best, and the worst performance over five independent runs are plotted in Figs. 4.14, 4.15, 4.16, and 4.17, respectively.

As shown in Fig. 4.14, it is straightforward to see that the learning performance is insensitive to the number of clients, especially for the experiments without applying partial DP. This makes sense since non-parametric models trained in VFL have nearly the same performance as those trained in the standard centralized learning [78]. And the experimental results without partial DP often have better performance than those with partial DP. However, by contrast, it is surprising to find that the root mean squared error (RMSE) of the Energy dataset with DP is slightly higher than those without DP across different numbers of clients. The reason behind this phenomenon might be that the model has overfit the training data without applying partial DP. In general, the performances with or without DP are almost the same and the test performance degradation resulting from the DP is almost negligible. Note also that the performance of the proposed PIVODL algorithm is rather stable in different runs.

Figure 4.15 shows the model performance over different maximum numbers of depth of the boosting tree. The test accuracy of the Credit card and Bank marketing datasets are around 82% and 90%, respectively. And these two curves remain almost constant as the number of maximum depth increases, except that the results of Credit card dataset with DP have slight fluctuations when the number of maximum depth is equal to 5. In addition, the model performances with or without DP are almost the same, indicating that the effect of partial DP is insensitive to the maximum depth of the tree. For the Energy dataset, the average RMSE decreases as the maximum depth increases, since more complex tree structures may have better model performance.

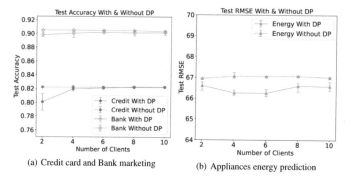

(a) Credit card and Bank marketing   (b) Appliances energy prediction

Fig. 4.14 The test accuracy and RMSE with and without applying DP over different numbers of participating clients, where **a** is the test accuracy on Credit card and Bank marketing datasets, and **b** is the test RMSE on the regression dataset

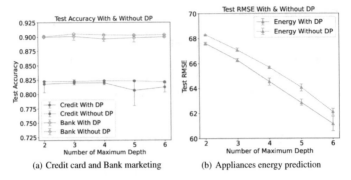

Fig. 4.15 The test accuracy and RMSE with or without applying DP over different maximum depths of a boosting tree, where **a** is the test accuracy on the Credit card and Bank marketing datasets, and **b** is the test RMSE on the regression dataset

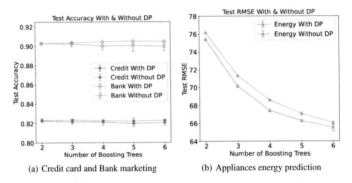

Fig. 4.16 The test accuracy and RMSE with or without applying DP over different numbers of boosting trees, where **a** is the test accuracy on the Credit card and Bank marketing datasets, and **b** is the test RMSE of the Appliances energy prediction dataset

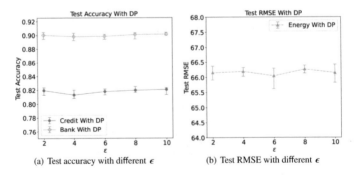

Fig. 4.17 The test performance over different ϵ values, where **a** is the test accuracy on the Credit card and Bank marketing datasets, and **b** is the test RMSE on Appliances energy prediction dataset

The performance variations over different numbers of boosting trees are shown in Fig. 4.16. The results of two classification tasks imply that the model performance has very slight changes as the number of boosting trees increases. On the Appliances energy prediction dataset, the RMSE slightly decreases with the increase in the number of boosting trees and the RMSE values with or without DP are nearly the same.

Surprisingly, the test performances of all three datasets also have no clear variations with the decrease of ϵ value (adding larger noise), as shown in Fig. 4.17. The possible reason behind this phenomenon is that partial DP is only applied on parts of predicted labels sent to the source client, significantly reducing the prediction update biases. Moreover, the split clients use de-noised leaf values for federated inference, which further reduces the performance biases caused by partial DP.

4.3.5.3 Evaluation on Training Time

Here, we empirically assess the training time with or without HE affected by different numbers of participating clients, maximum tree depth and boosting trees. All experiments are run on Intel Core i7-8700 CPU and the Paillier encryption together with the real number encoding strategy is implemented by the Python package in [79]. Note that the partial DP is not applied here, since the time consumption of DP is almost negligible.

At first, the training time with different numbers of clients are shown in Fig. 4.18. For the cases without using encryption, the training time of all three datasets almost keeps constant for different numbers of clients, which is consistent with the results reported in [67, 68]. This is because the training data samples are divided across their features in the VFL experiments, and the total number of participating clients has no effect on the entire data entries. On the other hand, the training time with the Paillier encryption increases linearly with the increase in the client numbers, since the amount of encryption times is proportional to the number of clients in PIVODL.

The training time over different maximum depths of a boosting tree grows exponentially without encryption, as shown in Fig. 4.19. This becomes more apparent when the Paillier encryption is applied due to $O(2^n)$ (n is the maximum depth of one decision tree) computational complexity of one decision. More specifically, the time consumption of the Appliance energy prediction dataset for 6-depth boosting trees are approximately 2 and 5 minutes without or with encryption, respectively.

The time consumption over different numbers of boosting trees increase linearly, as shown in Fig. 4.20. Similar to the previous results, the training time with the Paillier encryption is much larger than that without encryption. For example, the runtime on the energy dataset for six boosting trees is approximately 800 seconds with encryption, while it takes only about 350 seconds without encryption.

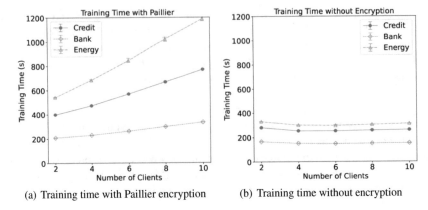

Fig. 4.18 The total training time over different numbers of participating clients, where **a** is the training time with Paillier encryption and **b** is the training time without encryption

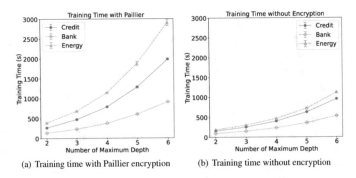

Fig. 4.19 The total training time over different maximum depths of a boosting tree, where **a** is the training time with Paillier encryption and **b** is the training time without encryption

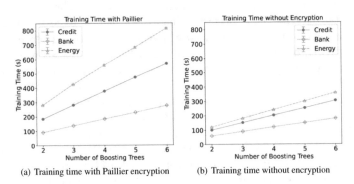

Fig. 4.20 The total training time over different number of boosting trees, where **a** is the training time with Paillier encryption and **b** is the training time without encryption

4.3.5.4   Evaluation of Communication Costs

In this section, we empirically analyze the communication costs (in MB) affected by the number of participating clients, the number of the maximum depth and the number of boosting trees.

Communication costs are evaluated by varying the number of clients for both cases with or without using the Paillier encryption. As shown in Fig. 4.21, the total communication costs for all three datasets have a rapid growth when the number of clients increases from two to four. The increase becomes relatively less fast when the number of clients further increases from four to ten, for both cases with and without the Paillier encryption. We can see that the communication costs increase with the Paillier encryption, but not as much as the training time compared to those without HE. The reason is that only encrypted gradient and Hessian sums of each possible split other than encrypted gradients and Hessians of each data point are transmitted among clients. Therefore, the communication costs are proportional to the number of buckets and is independent of the data size.

The communication costs related to different maximum tree depths are plotted in Fig. 4.22, from which we see that the communication costs are proportional to the the the maximum tree depth in both encryption and non-encryption scenarios. The communication costs of the Appliances energy prediction dataset are the largest among all three datasets for 290 MB without encryption and 425 MB with encryption for a depth of six.

The communication costs over the number of boosting trees are shown in Fig. 4.23. Similar to the previous results, the communication costs increase linearly with the number of trees, since the model size of XGBoost is proportional to the number of trees. And it is clear to see that the impact of encryption on communication costs is not as significant as that on the training time. The encryption brings a maximum of 25MB extra communication costs for the model with six boosting trees on the Appliance energy prediction dataset.

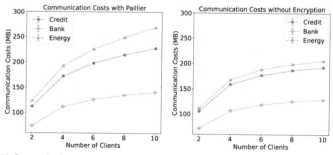

(a) Communication costs with Paillier encryp- (b) Communication costs without encryption
tion

Fig. 4.21 The total communication costs over different number of clients, **a** with the Paillier encryption, and **b** without encryption

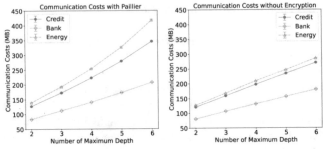

(a) Communication costs with Paillier encryp- (b) Communication costs without encryption
tion

Fig. 4.22 The total communication costs over different maximum depths of a boosting tree, where **a** is the communication costs with the Paillier encryption and **b** is the communication costs without encryption

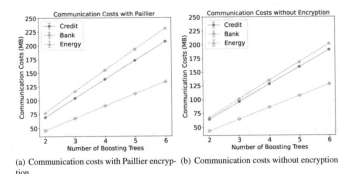

(a) Communication costs with Paillier encryp- (b) Communication costs without encryption
tion

Fig. 4.23 The total communication costs over different numbers of boosting trees, where **a** is the communication costs with the Paillier encryption and **b** is the communication costs without encryption

4.3.5.5 Inference of Predicted Labels

As introduced in Theorem 4.3, each client may deduce the true data labels from other clients through accumulated leaf values by simply integrating all received leaf values of a specific data entry and guessing its data label. So in this part, we compare and analyse the (average) guest accuracy of each participating client with and without partial DP.

To characterize the protection impact of the partial DP, we define the following guess accuracy:

Definition (*Guess Accuracy*) The accuracy of the attackers guessing the correct labels based on received leaf values.   □

It should be emphasized that the data labels are either 0 or 1 for binary classification problems, and the probability of a correct random guess is 50%. Therefore, as long

Table 4.5 Guess accuracy on different datasets

Dataset	Guess accuracy,no DP (%)	Guess accuracy, $\epsilon = 2$ (%)	Guess accuracy, $\epsilon = 4$ (%)	Guess accuracy, $\epsilon = 6$ (%)	Guess accuracy, $\epsilon = 8$ (%)	Guess accuracy, $\epsilon = 10$ (%)
Credit card	68.75	30.02	23.02	34.15	27.78	40.93
Bank marketing	61.57	39.57	38.85	25.85	39.73	14.45

as the guess accuracy is equal to or less than 50%, the label privacy will be regarded as successfully protected.

The results of two classification datasets with and without partial DP using different ϵ values are listed in Table 4.5. It is clear to see that the average guess accuracy for both two datasets are more than 60% without applying the partial DP, which is more accurate than a random guess. After applying the proposed partial DP algorithm, however, the guess accuracy drop dramatically and are all less than 50%. Specifically on the Credit card dataset, the guess accuracy decreases to around 27% for $\epsilon = 10$ and 45% for $\epsilon = 4$. On the Bank marketing dataset, the guess accuracy drops to 14.45% when $\epsilon = 10$. It is surprising to see that smaller ϵ values (adding larger noise) will give higher guess accuracy. The reason behind this is that some previous wrong guesses may be changed to correct guesses by adding undeterministic Gaussian noise. And as long as the guess accuracy is less than 50%, the label privacy is considered to be not revealed and the absolute accuracy does not really indicate the privacy preservation performance.

4.4 Summary

In this Chapter, we introduce two secure FL frameworks for both HFL and VFL. It is clear to see that privacy preservation mechanisms between HFL and VFL are extremely different. For secure HFL, most HE-based systems require a trusted third party (TTP) to generate key pairs for gradients encryption and decryption which deteriorates the topological structure of the standard FL. The proposed DAEQ-FL scheme distributedly generates key pairs between the central server and connected clients without the help of TTP. In addition, TernGrad and approximate aggregation techniques are adopted to simultaneously reduce both computational and communication costs. For secure VFL, most learning systems, however, do not contain a central server for model aggregation and the intermediate outputs other than the model gradients are encrypted before transmission. Most VFL systems assume that only one guest client contains all data labels, which is not realistic in the real world applications. Therefore, a secure VFL scheme called PIVODL is proposed, where a non-parametric model XGBoost is privately constructed with data labels distributed on multiple clients. Furthermore, a partial DP mechanism is proposed, which can not only prevent label inference leakage caused by prediction updates but also improve the performance of federated inference.

However, both of the two secure FL frameworks have very complex secure protocols and it is not easy to ensure all the steps strictly conducted according to the designed protocol in the real-world applications. In addition, frequent exchanges of Diffle-Hellman key or ciphertexts between clients make system management much more difficult. Consequently, more research work should be done to develop a lightweight and efficient secure FL system in the future.

References

1. McMahan, B., Moore, E., Ramage, D., Hampson, S., Arcas, B.A.y.: Communication-efficient learning of deep networks from decentralized data. In: Singh, A., Zhu, J. (eds.) Proceedings of the 20th International Conference on Artificial Intelligence and Statistics, Proceedings of Machine Learning Research, vol. 54, pp. 1273–1282. PMLR (2017). https://proceedings.mlr.press/v54/mcmahan17a.html
2. Lim, W.Y.B., Luong, N.C., Hoang, D.T., Jiao, Y., Liang, Y.C., Yang, Q., Niyato, D., Miao, C.: Federated learning in mobile edge networks: a comprehensive survey. IEEE Commun. Surv. Tutor. **22**(3), 2031–2063 (2020). https://doi.org/10.1109/COMST.2020.2986024
3. Luo, B., Li, X., Wang, S., Huang, J., Tassiulas, L.: Cost-Effective Federated Learning in Mobile Edge Networks (2021)
4. Hard, A., Rao, K., Mathews, R., Ramaswamy, S., Beaufays, F., Augenstein, S., Eichner, H., Kiddon, C., Ramage, D.: Federated Learning for Mobile Keyboard Prediction (2019)
5. Stremmel, J., Singh, A.: Pretraining federated text models for next word prediction. In: Arai, K. (ed.) Advances in Information and Communication, pp. 477–488. Springer International Publishing, Cham (2021)
6. Liang, X., Liu, Y., Luo, J., He, Y., Chen, T., Yang, Q.: Self-supervised cross-silo federated neural architecture search (2021)
7. Sheller, M.J., Edwards, B., Reina, G.A., Martin, J., Pati, S., Kotrotsou, A., Milchenko, M., Xu, W., Marcus, D., Colen, R.R., et al.: Federated learning in medicine: facilitating multi-institutional collaborations without sharing patient data. Sci. Rep. **10**(1), 1–12 (2020)
8. Rieke, N., Hancox, J., Li, W., Milletari, F., Roth, H.R., Albarqouni, S., Bakas, S., Galtier, M.N., Landman, B.A., Maier-Hein, K., et al.: The future of digital health with federated learning. NPJ Digit. Med. **3**(1), 1–7 (2020)
9. Yan, Z., Wicaksana, J., Wang, Z., Yang, X., Cheng, K.T.: Variation-aware federated learning with multi-source decentralized medical image data. IEEE J. Biomed. Health Inform. **25**(7), 2615–2628 (2021). https://doi.org/10.1109/JBHI.2020.3040015
10. Choudhury, O., Gkoulalas-Divanis, A., Salonidis, T., Sylla, I., Park, Y., Hsu, G., Das, A.: Differential privacy-enabled federated learning for sensitive health data (2020)
11. Li, X., Gu, Y., Dvornek, N., Staib, L.H., Ventola, P., Duncan, J.S.: Multi-site fmri analysis using privacy-preserving federated learning and domain adaptation: abide results. Med. Image Anal. **65**, 101765 (2020). https://doi.org/10.1016/j.media.2020.101765. www.sciencedirect.com/science/article/pii/S1361841520301298
12. Silva, S., Gutman, B.A., Romero, E., Thompson, P.M., Altmann, A., Lorenzi, M.: Federated learning in distributed medical databases: Meta-analysis of large-scale subcortical brain data. In: 2019 IEEE 16th International Symposium on Biomedical Imaging (ISBI 2019), pp. 270–274 (2019). https://doi.org/10.1109/ISBI.2019.8759317
13. Yan, B., Wang, J., Cheng, J., Zhou, Y., Zhang, Y., Yang, Y., Liu, L., Zhao, H., Wang, C., Liu, B.: Experiments of federated learning for covid-19 chest x-ray images. In: Sun, X., Zhang, X.,

Xia, Z., Bertino, E. (eds.) Advances in Artificial Intelligence and Security, pp. 41–53. Springer International Publishing, Cham (2021)

14. Guo, P., Wang, P., Zhou, J., Jiang, S., Patel, V.M.: Multi-institutional collaborations for improving deep learning-based magnetic resonance image reconstruction using federated learning. In: Proceedings of the IEEE/CVF Conference on Computer Vision and Pattern Recognition (CVPR), pp. 2423–2432 (2021)

15. Brisimi, T.S., Chen, R., Mela, T., Olshevsky, A., Paschalidis, I.C., Shi, W.: Federated learning of predictive models from federated electronic health records. Int. J. Med. Inform. **112**, 59–67 (2018). https://doi.org/10.1016/j.ijmedinf.2018.01.007. www.sciencedirect.com/science/article/pii/S138650561830008X

16. Lu, M.Y., Chen, R.J., Kong, D., Lipkova, J., Singh, R., Williamson, D.F., Chen, T.Y., Mahmood, F.: Federated learning for computational pathology on gigapixel whole slide images. Med. Image Anal. 102298 (2021). https://doi.org/10.1016/j.media.2021.102298, https://www.sciencedirect.com/science/article/pii/S1361841521003431

17. Li, H., Han, T.: An end-to-end encrypted neural network for gradient updates transmission in federated learning (2019)

18. Geiping, J., Bauermeister, H., Dröge, H., Moeller, M.: Inverting gradients—how easy is it to break privacy in federated learning? (2020)

19. Phong, L.T., Aono, Y., Hayashi, T., Wang, L., Moriai, S.: Privacy-preserving deep learning via additively homomorphic encryption. IEEE Trans. Inf. Forensics Secur. **13**(5), 1333–1345 (2018). https://doi.org/10.1109/TIFS.2017.2787987

20. Orekondy, T., Oh, S.J., Zhang, Y., Schiele, B., Fritz, M.: Gradient-leaks: Understanding and controlling deanonymization in federated learning (2020)

21. Shokri, R., Shmatikov, V.: Privacy-preserving deep learning. In: Proceedings of the 22nd ACM SIGSAC Conference on Computer and Communications Security, CCS '15, pp. 1310–1321. Association for Computing Machinery, New York, USA (2015). https://doi.org/10.1145/2810103.2813687, https://doi.org/10.1145/2810103.2813687

22. Cai, Z., Xiong, Z., Xu, H., Wang, P., Li, W., Pan, Y.: Generative adversarial networks: a survey toward private and secure applications. ACM Comput. Surv. **54**(6) (2021). https://doi.org/10.1145/3459992, https://doi.org/10.1145/3459992

23. Hitaj, B., Ateniese, G., Perez-Cruz, F.: Deep models under the gan: Information leakage from collaborative deep learning. In: Proceedings of the 2017 ACM SIGSAC Conference on Computer and Communications Security, CCS '17, pp. 603–618. Association for Computing Machinery, New York, USA (2017). https://doi.org/10.1145/3133956.3134012, https://doi.org/10.1145/3133956.3134012

24. Goodfellow, I., Pouget-Abadie, J., Mirza, M., Xu, B., Warde-Farley, D., Ozair, S., Courville, A., Bengio, Y.: Generative adversarial nets. In: Ghahramani, Z., Welling, M., Cortes, C., Lawrence, N., Weinberger, K.Q. (eds.) Advances in Neural Information Processing Systems, vol. 27. Curran Associates, Inc. (2014). https://proceedings.neurips.cc/paper/2014/file/5ca3e9b122f61f8f06494c97b1afccf3-Paper.pdf

25. Hardy, C., Le Merrer, E., Sericola, B.: Md-gan: multi-discriminator generative adversarial networks for distributed datasets. In: 2019 IEEE International Parallel and Distributed Processing Symposium (IPDPS), pp. 866–877 (2019). https://doi.org/10.1109/IPDPS.2019.00095

26. Wang, Z., Song, M., Zhang, Z., Song, Y., Wang, Q., Qi, H.: Beyond inferring class representatives: user-level privacy leakage from federated learning. In: IEEE INFOCOM 2019—IEEE Conference on Computer Communications, pp. 2512–2520 (2019). https://doi.org/10.1109/INFOCOM.2019.8737416

27. Zhang, J., Zhang, J., Chen, J., Yu, S.: Gan enhanced membership inference: a passive local attack in federated learning. In: ICC 2020—2020 IEEE International Conference on Communications (ICC), pp. 1–6 (2020). https://doi.org/10.1109/ICC40277.2020.9148790

28. Geiping, J., Bauermeister, H., Dröge, H., Moeller, M.: Inverting gradients—how easy is it to break privacy in federated learning? In: Larochelle, H., Ranzato, M., Hadsell, R., Balcan, M.F., Lin, H. (eds.) Advances in Neural Information Processing Systems, vol. 33, pp. 16937–16947. Curran Associates, Inc. (2020). https://proceedings.neurips.cc/paper/2020/file/c4ede56bbd98819ae6112b20ac6bf145-Paper.pdf

29. Zhu, L., Liu, Z., Han, S.: Deep leakage from gradients. In: Wallach, H., Larochelle, H., Beygelz-imer, A., d' Alché-Buc, F., Fox, E., Garnett, R. (eds.) Advances in Neural Information Processing Systems, vol. 32. Curran Associates, Inc. (2019). https://proceedings.neurips.cc/paper/2019/file/60a6c4002cc7b29142def8871531281a-Paper.pdf

30. Lim, J.Q., Chan, C.S.: From gradient leakage to adversarial attacks in federated learning. In: 2021 IEEE International Conference on Image Processing (ICIP), pp. 3602–3606 (2021). https://doi.org/10.1109/ICIP42928.2021.9506589

31. Huang, Y., Gupta, S., Song, Z., Li, K., Arora, S.: Evaluating gradient inversion attacks and defenses in federated learning (2021)

32. Dwork, C.: Differential privacy: a survey of results. In: Agrawal, M., Du, D., Duan, Z., Li, A. (eds.) Theory and Applications of Models of Computation, pp. 1–19. Springer, Berlin (2008)

33. Goldreich, O.: Secure multi-party computation. Manuscript. Preliminary version **78** (1998)

34. Gentry, C.: A Fully Homomorphic Encryption Scheme. Stanford university (2009)

35. Paillier, P.: Public-key cryptosystems based on composite degree residuosity classes. In: Stern, J. (ed.) Advances in Cryptology—EUROCRYPT '99, pp. 223–238. Springer, Berlin (1999)

36. Truex, S., Baracaldo, N., Anwar, A., Steinke, T., Ludwig, H., Zhang, R., Zhou, Y.: A hybrid approach to privacy-preserving federated learning. In: Proceedings of the 12th ACM Workshop on Artificial Intelligence and Security, AISec'19, pp. 1–11. Association for Computing Machinery, New York, USA (2019). https://doi.org/10.1145/3338501.3357370, https://doi.org/10.1145/3338501.3357370

37. Mandal, K., Gong, G.: Privfl: Practical privacy-preserving federated regressions on high-dimensional data over mobile networks. In: Proceedings of the 2019 ACM SIGSAC Conference on Cloud Computing Security Workshop, CCSW'19, pp. 57–68. Association for Computing Machinery, New York, USA (2019). https://doi.org/10.1145/3338466.3358926, https://doi.org/10.1145/3338466.3358926

38. Hao, M., Li, H., Xu, G., Liu, S., Yang, H.: Towards efficient and privacy-preserving federated deep learning. In: ICC 2019—2019 IEEE International Conference on Communications (ICC), pp. 1–6 (2019). https://doi.org/10.1109/ICC.2019.8761267

39. Zhang, C., Li, S., Xia, J., Wang, W., Yan, F., Liu, Y.: BatchCrypt: efficient homomorphic encryption for Cross-Silo federated learning. In: 2020 USENIX Annual Technical Conference (USENIX ATC 20), pp. 493–506. USENIX Association (2020). https://www.usenix.org/conference/atc20/presentation/zhang-chengliang

40. Zissis, D., Lekkas, D., Koutsabasis, P.: Cryptographic dysfunctionality-a survey on user perceptions of digital certificates. In: Georgiadis, C.K., Jahankhani, H., Pimenidis, E., Bashroush, R., Al-Nemrat, A. (eds.) Global Security, Safety and Sustainability and e-Democracy, pp. 80–87. Springer, Berlin (2012)

41. Postma, A., De Boer, W., Helme, A., Smit, G.: Distributed encryption and decryption algorithms. Memoranda Informatica 96–20 (1996)

42. Rivest, R.L., Shamir, A., Adleman, L.: A method for obtaining digital signatures and public-key cryptosystems. Commun. ACM **21**(2), 120–126 (1978). https://doi.org/10.1145/359340.359342

43. Agrawal, S., Mohassel, P., Mukherjee, P., Rindal, P.: Dise: Distributed symmetric-key encryption. In: Proceedings of the 2018 ACM SIGSAC Conference on Computer and Communications Security, CCS '18, pp. 1993–2010. Association for Computing Machinery, New York, USA (2018). https://doi.org/10.1145/3243734.3243774, https://doi.org/10.1145/3243734.3243774

44. Li, J., Huang, H.: Faster secure data mining via distributed homomorphic encryption. In: Proceedings of the 26th ACM SIGKDD International Conference on Knowledge Discovery & Data Mining, KDD '20, p. 2706-2714. Association for Computing Machinery, New York, USA (2020). https://doi.org/10.1145/3394486.3403321, https://doi.org/10.1145/3394486.3403321

45. Qiu, G., Gui, X., Zhao, Y.: Privacy-preserving linear regression on distributed data by homomorphic encryption and data masking. IEEE Access **8**, 107601–107613 (2020). https://doi.org/10.1109/ACCESS.2020.3000764

46. Aono, Y., Hayashi, T., Phong, L.T., Wang, L.: Privacy-preserving logistic regression with distributed data sources via homomorphic encryption. IEICE Trans. Inf. Syst. **99**(8), 2079–2089 (2016)

47. Wright, R.E.: Logistic Regression (1995)
48. Gennaro, R., Jarecki, S., Krawczyk, H., Rabin, T.: Secure distributed key generation for discrete-log based cryptosystems. In: Stern, J. (ed.) Advances in Cryptology—EUROCRYPT '99, pp. 295–310. Springer, Berlin (1999)
49. Shamir, A.: How to share a secret. Commun. ACM **22**(11), 612–613 (1979). https://doi.org/10.1145/359168.359176
50. Elgamal, T.: A public key cryptosystem and a signature scheme based on discrete logarithms. IEEE Trans. Inf. Theory **31**(4), 469–472 (1985). https://doi.org/10.1109/TIT.1985.1057074
51. Zhang, J., Li, M., Zeng, S., Xie, B., Zhao, D.: A survey on security and privacy threats to federated learning. In: 2021 International Conference on Networking and Network Applications (NaNA), pp. 319–326 (2021). https://doi.org/10.1109/NaNA53684.2021.00062
52. Mothukuri, V., Parizi, R.M., Pouriyeh, S., Huang, Y., Dehghantanha, A., Srivastava, G.: A survey on security and privacy of federated learning. Futur. Gener. Comput. Syst. **115**, 619–640 (2021). https://doi.org/10.1016/j.future.2020.10.007. www.sciencedirect.com/science/article/pii/S0167739X20329848
53. Pedersen, T.P.: Non-interactive and information-theoretic secure verifiable secret sharing. In: CRYTO. Springer, Berlin (1991)
54. Berrut, J.P., Trefethen, L.N.: Barycentric lagrange interpolation. SIAM Rev. **46**(3), 501–517 (2004)
55. ElGamal, T.: A public key cryptosystem and a signature scheme based on discrete logarithms. IEEE Trans. Inf. Theory **31**(4), 469–472 (1985)
56. Cramer, R., Gennaro, R., Schoenmakers, B.: A secure and optimally efficient multi-authority election scheme. Eur. Trans. Telecommun. **8**(5), 481–490 (1997)
57. Wen, W., Xu, C., Yan, F., Wu, C., Wang, Y., Chen, Y., Li, H.: Terngrad: Ternary gradients to reduce communication in distributed deep learning. In: Guyon, I., Luxburg, U.V., Bengio, S., Wallach, H., Fergus, R., Vishwanathan, S., Garnett, R. (eds.) Advances in Neural Information Processing Systems, vol. 30. Curran Associates, Inc. (2017). https://proceedings.neurips.cc/paper/2017/file/89fcd07f20b6785b92134bd6c1d0fa42-Paper.pdf
58. Uspensky, J.V.: Introduction to Mathematical Probability (1937)
59. Tarditi, D., Puri, S., Oglesby, J.: Accelerator: Using data parallelism to program gpus for general-purpose uses. SIGPLAN Not. **41**(11), 325–335 (2006). https://doi.org/10.1145/1168918.1168898
60. Zhao, B., Mopuri, K.R., Bilen, H.: idlg: Improved deep leakage from gradients (2020). arXiv:2001.02610
61. Fredrikson, M., Lantz, E., Jha, S., Lin, S., Page, D., Ristenpart, T.: Privacy in pharmacogenetics: an end-to-end case study of personalized warfarin dosing. In: 23rd {USENIX} Security Symposium ({USENIX} Security 14), pp. 17–32 (2014)
62. Fredrikson, M., Jha, S., Ristenpart, T.: Model inversion attacks that exploit confidence information and basic countermeasures. In: the 22nd ACM CCS
63. Wang, Z., Song, M., Zhang, Z., Song, Y., Wang, Q., Qi, H.: Beyond inferring class representatives: User-level privacy leakage from federated learning. In: IEEE INFOCOM 2019-IEEE Conference on Computer Communications, pp. 2512–2520. IEEE (2019)
64. Hitaj, B., Ateniese, G., Perez-Cruz, F.: Deep models under the gan: information leakage from collaborative deep learning. In: Proceedings of the 2017 ACM SIGSAC Conference on Computer and Communications Security, pp. 603–618 (2017)
65. Caldas, S., Wu, P., Li, T., Konecný, J., McMahan, H.B., Smith, V., Talwalkar, A.: LEAF: A benchmark for federated settings. CoRR **abs/1812.01097** (2018). http://arxiv.org/abs/1812.01097
66. Chen, T., Guestrin, C.: Xgboost: A scalable tree boosting system. In: Proceedings of the 22nd ACM SIGKDD International Conference on Knowledge Discovery and Data Mining, KDD '16, pp. 785–794. Association for Computing Machinery, New York, USA (2016). https://doi.org/10.1145/2939672.2939785, https://doi.org/10.1145/2939672.2939785
67. Cheng, K., Fan, T., Jin, Y., Liu, Y., Chen, T., Papadopoulos, D., Yang, Q.: Secureboost: a lossless federated learning framework. IEEE Intell. Syst. **36**(6), 87–98 (2021). https://doi.org/10.1109/MIS.2021.3082561

68. Tian, Z., Zhang, R., Hou, X., Liu, J., Ren, K.: Federboost: private federated learning for gbdt (2020). arXiv:2011.02796
69. Bonawitz, K., Ivanov, V., Kreuter, B., Marcedone, A., McMahan, H.B., Patel, S., Ramage, D., Segal, A., Seth, K.: Practical secure aggregation for privacy-preserving machine learning. In: Proceedings of the 2017 ACM SIGSAC Conference on Computer and Communications Security, CCS '17, pp. 1175–1191. Association for Computing Machinery, New York, USA (2017). https://doi.org/10.1145/3133956.3133982, https://doi.org/10.1145/3133956.3133982
70. Dwork, C., McSherry, F., Nissim, K., Smith, A.: Calibrating noise to sensitivity in private data analysis. In: Theory of Cryptography Conference, pp. 265–284. Springer, Berlin (2006)
71. Dwork, C., Kenthapadi, K., McSherry, F., Mironov, I., Naor, M.: Our data, ourselves: Privacy via distributed noise generation. In: Annual International Conference on the Theory and Applications of Cryptographic Techniques, pp. 486–503. Springer, Berlin (2006)
72. Abadi, M., Chu, A., Goodfellow, I., McMahan, H.B., Mironov, I., Talwar, K., Zhang, L.: Deep learning with differential privacy. In: Proceedings of the 2016 ACM SIGSAC Conference on Computer and Communications Security, CCS '16, pp. 308–318. Association for Computing Machinery, New York, USA (2016). https://doi.org/10.1145/2976749.2978318, https://doi.org/10.1145/2976749.2978318
73. Geyer, R.C., Klein, T., Nabi, M.: Differentially private federated learning: a client level perspective (2017). arXiv:1712.07557
74. Wei, K., Li, J., Ding, M., Ma, C., Yang, H.H., Farokhi, F., Jin, S., Quek, T.Q.S., Poor, H.V.: Federated learning with differential privacy: algorithms and performance analysis. IEEE Trans. Inf. Forensics Secur. **15**, 3454–3469 (2020). https://doi.org/10.1109/TIFS.2020.2988575
75. Yeh, I.C., Lien, C.h.: The comparisons of data mining techniques for the predictive accuracy of probability of default of credit card clients. Expert Syst. Appl. **36**(2), 2473–2480 (2009)
76. Moro, S., Cortez, P., Rita, P.: A data-driven approach to predict the success of bank telemarketing. Decis. Support. Syst. **62**, 22–31 (2014)
77. Candanedo, L.M., Feldheim, V., Deramaix, D.: Data driven prediction models of energy use of appliances in a low-energy house. Energy Build. **140**, 81–97 (2017)
78. Zhu, H., Xu, J., Liu, S., Jin, Y.: Federated learning on non-iid data: a survey. Neurocomputing **465**, 371–390 (2021). https://doi.org/10.1016/j.neucom.2021.07.098, www.sciencedirect.com/science/article/pii/S0925231221013254
79. Data61, C.: Python Paillier Library. https://github.com/data61/python-paillier (2013)

Chapter 5
Summary and Outlook

Abstract Each chapter should be preceded by an abstract (no more than 200 words) that summarizes the content. The abstract will appear *online* at www.SpringerLink. com and be available with unrestricted access. This allows unregistered users to read the abstract as a teaser for the complete chapter.

5.1 Summary

Federated learning is an emerging research area that has attracted increasing attention in the artificial intelligence research community and witnessed a wider range of real-world applications, including recommender systems, healthcare, financial technologies, human-machine interactions, smart cities, internet of things, and edge computing.

This book addresses the fundamental and most important challenges in federated learning. First, we present various ideas for reducing communication costs in federated learning. One method for communication-efficient federated learning is layer-wise asynchronous model update in combination of temporally weighted aggregation. This method is based on the assumption that not all layers in a deep neural network need to be updated in each communication round, since the deep and shallow layers the deep neural networks are responsible for learning different features. Another method for reducing communication costs is the use of ternary compression of the weights to be transmitted. To maintaining high learning performance while reducing the communication costs, a trained ternary compression method has been adopted. It is shown theoretically that one same coefficient is adequate for positive and negative weights. Another interesting finding is that ternary compression is helpful in alleviating model divergence in the presence of non-iid training data. Finally, communication-efficiency has also been accomplished by reducing the complexity of the deep learning models by means of neural architecture search, although neural architecture search can also enhance the performance, reduce the memory requirements, and improve the robustness of the neural network architecture.

Neural architecture search under the federated environment is the next main topic of the book, which becomes more challenging than in a centralized learning environment for several reasons. First, the influence of the non-iid data may becomes much stronger, since different local clients may find very different neural architectures, making model divergence more likely. Thus, a better idea for handling non-iid data must be developed. Second, the real-time performance becomes harder to achieve, when the computing capability of the local devices are limited and when population-based search methods such as evolutionary algorithms are employed. To address the above issues, a supernet is adopted in combination of a model sampling and client sampling technique. Model sampling based on the supernet and weight sharing makes it easier for model aggregation without taking additional strategies such as client clustering or personalization. Meanwhile, client sampling ensures that each client only needs to evaluate one candidate architecture at a time, thereby enhancing the real-time performance.

Although federated learning provides an efficient framework for protecting the privacy of the data, there is still a risk of leaking the data if there are adversarial users. To further enhance the security and privacy-preservation level, additional privacy computing techniques can be introduced on top of federated learning, such as differential privacy and homomorphic encryption. To improve the computational efficiency, we present a distributed encryption method for horizontal federated learning on the basis of ternary compression and approximate aggregation. Additionally, in a vertical federated learning setting, a method for secure node splitting and construction is developed, and a partial differential privacy algorithm is designed to preserve the sensitive data.

5.2   Future Directions

This book gives an introductory yet complete picture of federated learning, paving the way for researchers to further explore research challenges or practitioners to apply federated learning to real-world problems. Nevertheless, a few important points discussed below have not been treated in detail in this book.

- Quantitative definitions of privacy and security are needed. In most federated learning frameworks, threat models are defined regarding the server and clients. For example, it is assumed that the server is honest-but-curious, and the majority of the clients are not adversarial users. However, quantitative definitions are still missing, and privacy-preservation is typically treated as confidentiality. In fact, it may happen that some non-sensitive data can be given out, and the degree of sensitivity of different data attributes may be different. Simply treating all data as confidential may unnecessarily reduce the performance of the resulting machine models.
- Although additional techniques such as multi-party secure computing, differential privacy and homomorphic encryption can further enhance privacy protection, they are either computationally intensive or deteriorate the learning performance. One recent technique is to use the so-called trusted execution environments to provide a

hardware platform for secure execution of code and handling of data, considerably reducing training and communication time. Moreover, blochchain techniques can be integrated with trusted execution environments.

- Heterogeneity in the federated environment must be more carefully taken into account. By heterogeneity, we mean the differences in computational and communication capabilities of different devices. Thus, it makes sense to consider assign models of different complexities to different devices, and the asynchronous aggregation of different models. Another aspect of heterogeneity is the architecture of federated learning environment. At present, most federated learning assumes a flat structure, i.e., a server and a number of local devices. However, in practice, e.g., in an Internet of Things system, the computing architecture may contain several levels of computing hierarchies and multiple servers. In such a system, the relationship between different devices will be more complicated.
- Fairness and preferences of the local clients must be taken into account. On the one hand, fairness may mean to give different incentives to different users who contribute differently in the federated learning system. On the other hand, fairness may also mean that different users may have different preferences. Consequently, an optimal solution for the overall system may present biases to the individual users.

Index

Y. Jin et al., *Federated Learning*, Machine Learning: Foundations, Methodologies,
and Applications, https://doi.org/10.1007/978-981-19-7083-2_3

Printed in the United States
by Baker & Taylor Publisher Services